Healthcare Entrepreneurship and Management

Post pandemic, the world is not the same place. There has been an increasing focus on healthcare and well-being, which has created a once-in-a-lifetime opportunity for healthcare innovations and startups. From adoption of a range of medical apps and telemedicine technologies to heightened public interest in smart wearables and medical devices, the demand for efficient healthcare delivery has been skyrocketing.

This book aims to serve as a first-of-its-kind guide for skill development in conception to commercialisation of healthcare products and services. It covers the gamut from the study of healthcare challenges, such as understanding customer requirements, market needs, and competition, to the various steps of the healthcare product development process, such as defining value propositions and specifications, the creation of minimum viable product (MVP) to prototyping, and manufacturing. The authors also discuss key commercialisation and management strategies, including the development of a robust business plan, fund raising, intellectual property, creating barriers to entry, and launching healthcare startups. Medical product pricing, positioning, sales and distribution, and customer acquisition are also presented with real-life examples.

This book serves as a key reference not only for biomedical engineers who are looking to launch their products or services in the market but also for budding entrepreneurs willing to explore opportunities in the healthcare domain. For example, engineers and managers working on the development of medical devices require knowledge of ethical guidelines, regulations, and approvals to effectively launch their products in the medtech industry. On the other hand, entrepreneurs looking to benefit from the booming healthcare industry will find this book helpful in understanding the fundamentals of medical product development and commercialisation to launch their ideas successfully.

Healthcare Entrepreneurship and Management

A Comprehensive Guide for Biomedical Engineers and Entrepreneurs

Arnab Chanda
and Shubham Gupta

A PRODUCTIVITY PRESS BOOK

First published 2025
by Routledge
605 Third Avenue, New York, NY 10158

and by Routledge
4 Park Square, Milton Park, Abingdon, Oxon, OX14 4RN

Routledge is an imprint of the Taylor & Francis Group, an informa business

ISBN: 978-1-032-75709-4 (hbk)
ISBN: 978-1-032-75708-7 (pbk)
ISBN: 978-1-003-47530-9 (ebk)

DOI: 10.4324/9781003475309

Typeset in Garamond
by Deanta Global Publishing Services, Chennai, India

Contents

INTRODUCTION I

Chapter 1

Introduction to Current Healthcare Problems and Unmet Needs

1.1 Introduction

The lives of patients are significantly impacted by hospitals, community health clinics, and healthcare systems [1]. US healthcare providers and entrepreneurs can support patients in satisfying their unmet fundamental requirements while also educating them on the connections between health and unmet needs, or social determinants of health (SDOH). Although Medicaid-insured patients are disproportionately affected by unmet basic requirements, over 45% of consumers with all coverage categories report having at least one unmet basic need. It's possible that patients with increased incomes are nonetheless experiencing unmet requirements. Ninety percent of the $3.8 trillion that the United States spends annually on healthcare goes toward treating persons with mental health and chronic illnesses [2]; 60% of adult Americans have one chronic illness [3], and 40% have two or more [1].

Almost 25% of families in several low- and middle-income nations avoided medical attention when it was necessary during the COVID-19 pandemic due to financial constraints, movement restrictions, or fear of getting the virus [4]. Estimates indicate that nearly half of adult Europeans (aged 18–29 years old) experienced lack of fulfilment of needs pertaining to mental healthcare throughout the COVID-19 pandemic, even in high-income environments [5,6]. Unmet needs are those for which a person has a need

DOI: 10.4324/9781003475309-2

for healthcare but is unable or unwilling to obtain high-quality treatment [7]. Poor health outcomes, excessive spending, and a decline in both individual and societal productivity could result from this [8]. Unmet need is not taken into account in the present measures of the health system in providing services. The 76th World Health Assembly, 2023, recently started to examine the value and viability of adding unmet healthcare needs as a metric to track universal health coverage (UHC) both domestically and internationally [4].

Unmet healthcare need is exacerbated by unfulfilled fundamental needs, which also result in poor results, needless and unnecessary utilisation, and clinical exacerbations [9]. Contrasted with individuals whose fundamental needs are satisfied, those grappling with unmet basic requirements are considerably more inclined to express subpar physical well-being, mental health challenges, and a pronounced increase in healthcare usage—surpassing 2.5 times for physical well-being, fivefold for mental health, and doubling for healthcare utilisation [1]. Moreover, individuals facing unmet basic needs typically encounter restricted access to healthcare and express lower satisfaction with the care they do receive [10]. Even among those employed, those with one or more unmet basic needs face a 2.4-fold heightened risk of absenteeism, missing six or more days of work in the preceding year, and forgoing necessary medical treatment [1].

Closing the gaps in unmet requirements could boost economic output while also immediately improving patient and community care. Healthcare providers and entrepreneurs should think about taking certain steps, such as gathering patient data, to determine which unmet basic needs remedies are most likely to be successful [1]. In addition to this, a few systematic approaches for uncovering medical needs could be followed, such as firstly defining the need, understanding need statement and prioritising needs, defining need specification, and then following an algorithmic approach (Figure 1.1).

In order to attain optimal health outcomes, timely utilisation of personal health services is referred to as "access to health care" [12]. Individuals may face both non-financial and financial obstacles that limit their ability to

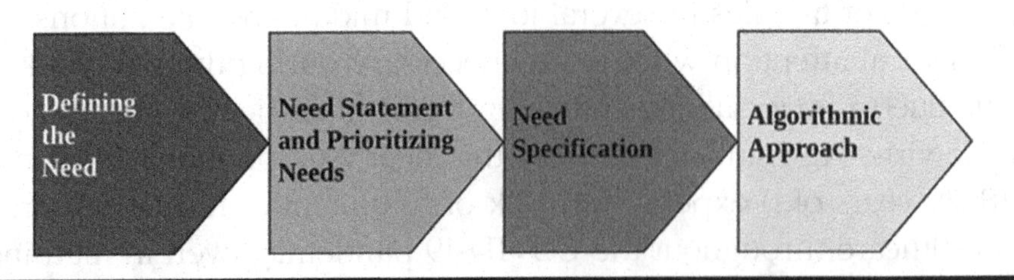

Figure 1.1 A methodical technique to identify medical requirements [11].

obtain the necessary medical treatment. Poorer health outcomes may arise from delayed or non-received care. The proportion of individuals across all age groups who postponed or refrained from receiving necessary medical care because of expense declined from 11.4% in 2009 to 7.2% in 2016, and subsequently rose to 8.3% in 2018. In 2023, 8.5% of people postponed or did not receive necessary medical care because of cost [12]. From 8.4% in 2009 to 5.2% in 2015, the percentage of people of all ages who did not receive necessary prescription drugs owing to cost in the previous 12 months declined, and it remained relatively stable until 2018 (5.3%) [12]. In addition to this, 5.6% of people in 2019 reported that they had not received necessary prescription medication in the previous 12 months owing to the cost [12]. Conversely, the proportion of individuals two years of age and older who did not receive necessary dental care within the previous year owing to expense declined from 13.3% in 2009 to 8.9% in 2016, and then rose to 9.9% in 2018. In 2019, 14.7% of people aged over two did not receive necessary dental care within the previous year due to financial constraints [12].

In 2022, specifically for EU (European Union) citizens, 4.1% of people aged 16 years or above reported that they had unfulfilled needs for a medical examination or treatment [13]; this percentage varied from 0.2% in Cyprus to 12.5% in Denmark, 12.6% in Estonia, and 13.1% in Greece. 2.2% of EU citizens reported having unmet needs, with the percentage varying from 0.1% in Cyprus to 9.0% in Greece and 9.1% in Estonia. These respondents cited factors such as cost, distance from place of residence, or waiting lists that affected the organisation and operation of healthcare services (Figure 1.2) [13]. In the 2023 survey, 46% of adults globally said that the largest issue confronting their nation's healthcare system was either a lack of treatment or lengthy wait times (Figure 1.3) [14]. An additional 46% of adults stated that the largest problem was a staffing shortage (Figure 1.3). This statistic displays the proportion of respondents worldwide who said that, as of 2023, a few specific concerns were the main challenges facing their nation's healthcare system [14].

1.2 Creating an Action Plan

Despite the initiation of efforts by healthcare entrepreneurs and providers to acknowledge unmet basic needs through screening, closed-loop referrals, resource managers, and community relationships, the transition from small-scale initiatives serving a specific patient subset to comprehensive programmes is met with formidable challenges. These obstacles encompass constrained resources, a

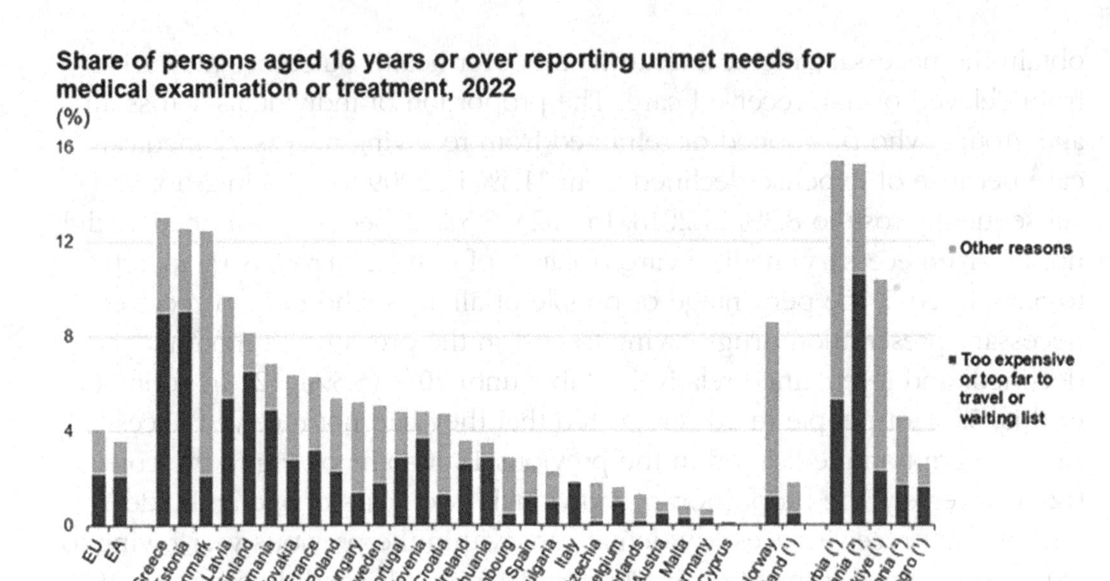

Figure 1.2 Share of persons aged 16 years or over reporting unmet needs for medical examination or treatment [13].

lack of data pinpointing who would benefit from the interventions, insufficient coordination to identify needs and implement scaled solutions across various sites in the hospital, medical departments, and programmes, along with uncertainties regarding the economic return on investment.

No healthcare practitioner or entrepreneur can solve every social issue, and they shouldn't feel compelled to. But healthcare, especially value-based care initiatives, may give priority to initiatives that would improve patient outcomes the most, fit in with existing or planned partnerships and capabilities, and be financially viable [15]. The steps below can be used by healthcare entrepreneurs or providers to create an efficient programme [1].

Considering unmet needs as strategic priorities: Determining how addressing patients' unmet basic needs can contribute to the realisation of the system's strategic objectives is crucial. Unattended basic requirements among patients exert a substantial influence on their engagement, experience, and overall health, emphasising the interconnectedness between addressing these needs and the attainment of broader organisational goals [16]. Additionally, they have a close relationship with institutional strategic targets like value-based care, quality improvement, patient experience, and

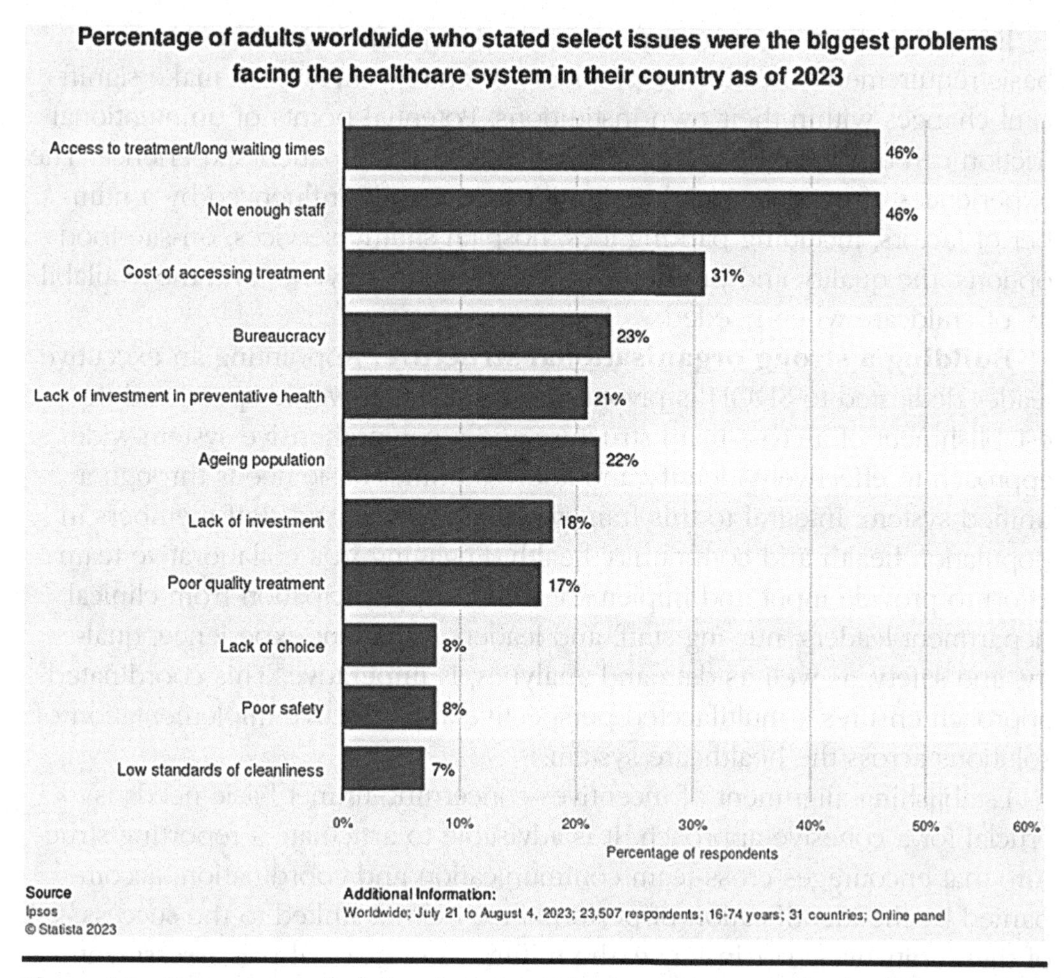

Figure 1.3 Percentage of adults who reported selected issues in the healthcare system [14].

the overall cost of care. Entrepreneurs in the healthcare industry or providers should begin by determining which unmet needs are most directly related to current strategic priorities and capabilities, then concentrating on those (for example, by communicating the link with other leaders across the organisation to gain traction and dedicate funding).

Unaddressed basic needs can be interconnected with various strategic priorities, and multiple avenues to address SDOH exist. Informed by insights into the impact of unmet basic needs on clinical outcomes, the prevalence of these needs in communities, existing organisational capabilities, and the potential for forming partnerships with community-based organisations or other healthcare entities, healthcare entrepreneurs or providers can strategically prioritise their initiatives and determine where to commence interventions.

Even though there are many external opportunities to address unmet basic requirements, healthcare providers or entrepreneurs can make significant changes within their own institutions. Potential points of unintentional friction can be found by taking a holistic view of the patient experience. The experiences of patients and their families are greatly influenced by a number of factors, including parking fees, hospital shuttle services, on-site food options, the quality and promptness of translation services, and the availability of childcare when needed.

Building a strong organisational structure: Appointing an executive leader dedicated to SDOH is pivotal. This individual would spearhead the establishment of a cross-team structure and a comprehensive system-wide approach to effectively identify and address unmet basic needs through a unified system. Integral to this framework are specialised staff members in population health and community health. Engaging in a collaborative team effort to provide input and implement solutions, participation from clinical department leaders, nursing staff, and leaders in patient experience, quality, and safety, as well as data and analytics, is imperative. This coordinated approach ensures a multifaceted perspective and effective implementation of solutions across the healthcare system.

Establishing alignment of incentives concerning unmet basic needs is crucial for a cohesive approach. It is advisable to articulate a reporting structure that encourages cross-team communication and coordination, accompanied by the identification of performance metrics linked to the success of these initiatives. For instance, the fulfilment of a social risk assessment could serve as a process metric for primary-care physicians operating within value-based-care contracts. Additionally, to comprehend patient perspectives consistently, cultivating robust relationships with patients and community members proves invaluable for programme development. Healthcare entrepreneurs or provider systems might consider establishing family and patient advisory councils and committees, ensuring active engagement with leaders throughout the hospital system.

Collection of data: Establishing priorities for data collection to enable particular activities. Healthcare providers or entrepreneurs should make sure that data-gathering initiatives help improve comprehension of the unmet basic demands and the programmes that now address those needs. For instance, a healthcare professional or entrepreneur offering a programme for non-emergency medical transportation could be aware of the individuals in need, their residences, and any other needs that the programme might be able to address (for example, grocery or prescription collection). In an ideal world, programme feedback, engagement, and correlation with health outcome measurements (e.g., improved control over a chronic health condition, stress reduction, decreased depression, or reduced anxiety) would all be routinely assessed and might guide future programme iterations.

Creating screening procedures that connect with a variety of patients at pivotal times is essential. Annual preventative checkups are increasingly often combined with checks for unmet basic needs. Nevertheless, there may be additional high-value times when this kind of screening takes place, like right before hospital discharge or during an ER visit. Furthermore, in order to avoid perceived stigma, some patients could prefer to self-screen and locate resources; others would prefer to do so on an as-needed basis rather than waiting for a visit from a healthcare professional or entrepreneur.

Businesses in the healthcare industry that offer vendor platforms with technology that enables self-screening and extensive resource libraries can help patients. They might also work with patients or direct them to local services. Basic requirements screening and billing procedures need to be straightforward and effective because frontline healthcare professionals and businesses are required to execute an increasing number of tasks and screenings during patient visits. Provider systems or healthcare entrepreneurs may be able to achieve this by including these screening technologies with patient appointments.

1.3 Creating a Support Infrastructure and Investigating Outcomes

Having solid ties with neighbourhood-based organisations may be advantageous for healthcare providers or enterprises. To improve the patient experience, they can try "warm handoffs" between social service and medical healthcare providers or entrepreneurs. It might be possible to automate this procedure by investing in social services referral infrastructure and technology. Through the establishment of referral networks facilitating communication between healthcare providers and social service providers, healthcare providers can verify whether patients have received the necessary services. This lays the groundwork for enhanced care coordination, allowing for a more comprehensive and integrated approach to addressing patients' needs.

In order to detect unmet basic requirements and connect patients to services, many hospitals are forming teams of coordinators and providing training to permanent staff, such as community health workers and patient navigators. Social workers are under less stress because of this referral system. Additionally, providers may wish to assist patients in understanding the benefits that are included in the healthcare plan. Furthermore, players broaden their range of offerings to include additional benefits, such as providing medically tailored meals for patients with specific health conditions.

The provider may decide on measures depending on possible outcomes. These objectives should ideally be connected to broad strategic key performance indicators (KPIs). For example, by working to standardise screening for unmet basic requirements and referrals, providers might monitor the proportion of screened patients and show if referrals resulted in a decrease in unmet basic needs. The provider may link these measurements to a broad KPI pertaining to other screenings. Following the establishment of metrics and realistic goals, the provider could adopt an agile or quality improvement approach to improvement by conducting iterative measurements and adjustments based on input from staff, patients, families, and community members.

1.4 Forming and Prioritising a Need Statement

Making a need statement—a succinct description of the invention's purpose—is the first step in inventing anything new. Clear definitions of the problem, population, and outcome are necessary for a need statement to be effective. Consider the following three need statements [11].

1. A strategy to avoid problems after sternotomy.
2. A strategy to stop sternotomy wire-related infections.
3. A non-sternotomy method of revascularising the coronaries.

While sternotomy staples might have been contemplated as a potential solution for the initial need statement, it is acknowledged that the statement lacks specificity, rendering it a suboptimal example of a clear and focused statement. Now, for the second need statement, the quest for a method to reduce infections offers additional insights in addition to the specificity into understanding the primary reasons for complications. In contrast, the third need statement expands our perspective by exploring methods for minimally invasive revascularisation of coronary arteries following sternotomy. The formulation of this third need statement, for instance, could have been a catalyst for the creation of percutaneous procedures such as stenting.

Addressing challenges in cancer therapy requires a nuanced consideration of various aspects. Several formulations of the need statement could be explored [11]:

1. Developing an Effective Cure for Cancer: Crafting methodologies aimed at discovering a definitive cure for cancer, considering diverse modalities and approaches.
2. Minimising Chemotherapy Side Effects while Preserving Anti-Tumour Efficacy: Innovating strategies to mitigate the adverse effects of chemotherapy, ensuring a balance that preserves the anti-tumour efficacy of the treatment.
3. Non-Invasive Ablation of Stage IA Non-Small Cell Lung Cancer (NSCLC) Tumours: Devising techniques for the non-invasive ablation of stage IA NSCLC tumours, eliminating the need for thoracotomy and advancing patient-friendly treatment options.

Each need statement above addresses a distinct aspect of cancer therapy, reflecting the multifaceted nature of the challenge and the diverse approaches required to enhance outcomes in cancer treatment.

The initial requirement merely outlines a broad objective without providing additional information or guidelines to guide the invention process [11]. Alternatively, develop a need statement along the lines of need statement 2, where the objective is to cure cancer with chemotherapy while reducing toxicity. A need statement with a well-defined objective, like the one described, could have been instrumental in inspiring the creation of the transarterial chemoembolisation (TACE) treatment, which currently stands as the gold standard for treating hepatocellular carcinoma. Additionally, the meticulously crafted third need statement holds the potential to drive the development of an invention for bronchoscopic ablation of stage IA NSCLC tumours. Such an innovation could substantially minimise the morbidity and mortality traditionally associated with conventional chest-opening treatments, providing a notably less intrusive alternative.

In order to enhance the invention's chances of success, it is crucial to prioritise creating well-crafted need statements, as inventions take time, resources, and funding. The size of the market, the gap in technology, and reimbursement are the three main factors that determine how to rank a need statement. For instance, a sizeable patient base with a more severe ailment will have a sizeable market, which might greatly enhance the quality of life for patients.

Launching an invention would be challenging if the patient group was small and the severity of the sickness was low. This would limit the possibility of reaching and significantly impacting those people. The technology gap is the next need statement to be taken into account. If there are many emerging technologies, the gap in technical advancement can be extremely tiny; conversely, if there is a shortage of newly produced technology, the gap can be very significant. The opportunity costs associated with releasing a new invention onto the market are represented by the extent of the technology gap. For instance, it will be harder to succeed in marketplaces where new technologies are widely available yet have little to no influence. Thirdly, the ability to recover the expenses incurred in creating and producing an innovation is essential to its success.

For the need specification, comprehensive data are needed once a specific requirement statement has been discovered. The need specifications cover every aspect needed for an invention to be commercialised, such as: 1) medical characteristics; 2) stakeholders; 3) legal requirements; 4) reimbursement; 5) possible acquirers; and 6) marketing [11]. The components of an innovation that is going to be sold are depicted in Figure 1.4. An inventor must evaluate the procedure's level of invasiveness and the clinical endpoints that need to be reached in terms of medical qualities. If a large-scale clinical trial is necessary to demonstrate the safety and efficacy, the

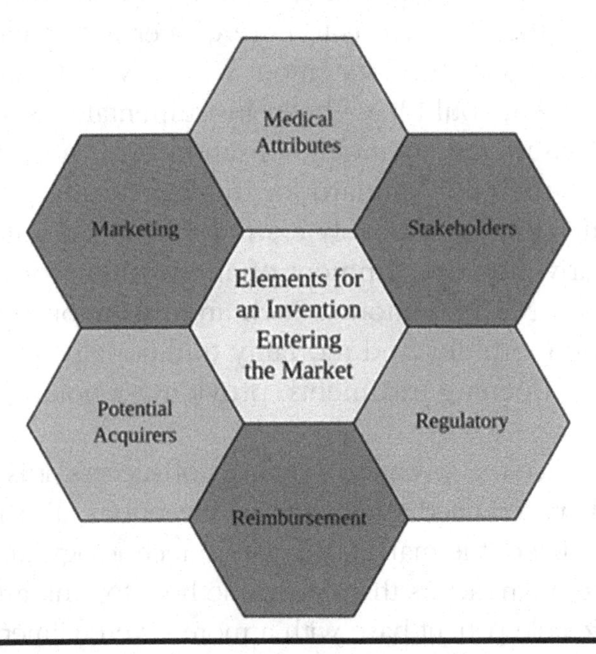

Figure 1.4 Elements for an invention entering the market [11].

accompanying costs and feasibility must be taken into account. Applicability and usability can be determined with the assistance of significant stakeholders, such as the doctors who will carry out the treatments. Regulatory agencies may also provide information about the size of a trial that is required to assess safety and efficacy and if a device needs premarket approval.

Finally, how well a product enters the market will depend on leads to possible acquirers and marketing tactics [11]. Venture capital companies and investors will need specific reimbursement pathways and market research. Members of the target audience are more likely to accept the device's implementation as a viable, workable solution if all requirements in the need specification are satisfied.

An inventor can concentrate on creating the novel device once the need statement and need specification have been met. For this stage, an algorithmic technique can be applied. To begin with, examine previous and present treatments to determine the reasons for each treatment's success and failure [11]. Secondly, turn a sickness to your advantage. Urinary retention mechanisms, for instance, can be mimicked in a urinary incontinence treatment device. Finally, consider how analogous processes arise in the natural world. Using these techniques will probably help inventors come up with a few ideas for new products that meet the given demand. Once these innovations are underway, compile the various requirements specifications for each invention and verify the intellectual property (IP) positions of the top ideas.

References

1. Patients struggle with social determinants of health: medical providers can help | McKinsey n.d. https://www.mckinsey.com/industries/healthcare/our-insights/patients-struggle-with-unmet-basic-needs-medical-providers-can-help (accessed January 15, 2024).
2. Health and economic costs of chronic diseases | CDC n.d. https://www.cdc.gov/chronicdisease/about/costs/index.htm (accessed January 15, 2024).
3. Chronic diseases in America | CDC n.d. https://www.cdc.gov/chronicdisease/resources/infographic/chronic-diseases.htm (accessed January 15, 2024).
4. Rosenberg M, Kowal P, Rahman MM, Okamoto S, Barber SL, Tangcharoensathien V. Better data on unmet healthcare need can strengthen global monitoring of universal health coverage. *BMJ* 2023;382:e075476. https://doi.org/10.1136/BMJ-2023-075476.
5. Kim JY, Kim DI, Park HY, Pak Y, Tran PNH, Thai TT, et al. Unmet healthcare needs and associated factors in rural and suburban Vietnam: a cross-sectional study. *Int J Environ Res Public Heal* 2020;17:6320. https://doi.org/10.3390/IJERPH17176320.

6. Yadav R, Yadav J, Shekhar C. Unmet need for treatment-seeking from public health facilities in India: An analysis of sociodemographic, regional and disease-wise variations. *PLoS Glob Public Heal* 2022;2:e0000148. https://doi.org/10.1371/JOURNAL.PGPH.0000148.
7. Academy of Medical Sciences. Unmet need in healthcare. n.d.
8. Health at a glance: Europe 2022 2022. https://doi.org/10.1787/507433B0-EN.
9. Chen J, Hou F. Unmet needs for health care. *Stat Canada* 2002;13:82–5.
10. Rahman MM, Rosenberg M, Flores G, Parsell N, Akter S, Alam MA, et al. A systematic review and meta-analysis of unmet needs for healthcare and long-term care among older people. *Health Econ Rev* 2022;12:1–10. https://doi.org/10.1186/S13561-022-00398-4/FIGURES/3.
11. Dao D, Brenner J. Identifying unmet needs: problems that need solutions. *Acad Entrep Med Heal Sci* 2019. https://doi.org/10.21428/B2E239DC.171B57D1.
12. Unmet need for health care - health, United States n.d. https://www.cdc.gov/nchs/hus/topics/unmet-need.htm (accessed January 15, 2024).
13. Unmet health care needs statistics - statistics explained n.d. https://ec.europa.eu/eurostat/statistics-explained/Unmet_health_care_needs_statistics (accessed January 15, 2024).
14. Leading problems in healthcare systems worldwide 2023 | Statista n.d. https://www.statista.com/statistics/917153/leading-problems-healthcare-systems-worldwide (accessed January 15, 2024).
15. The six biggest challenges of the hospital at home care model n.d. https://us.nttdata.com/en/blog/2023/december/the-six-biggest-challenges-of-the-hospital-at-home-care-model (accessed January 15, 2024).
16. Kowal P, Corso B, Anindya K, Andrade FCD, Giang TL, Guitierrez MTC, et al. Prevalence of unmet health care need in older adults in 83 countries: measuring progressing towards universal health coverage in the context of global population ageing. *Popul Health Metr* 2023;21:1–16. https://doi.org/10.1186/S12963-023-00308-8/FIGURES/6.

Chapter 2

Healthcare Startup Ecosystems around the Globe

2.1 Introduction

Industries may be disrupted and reshaped by the enormous forces created by ecosystems [1]. In the healthcare sector, they have the ability to boost customer satisfaction and affordability, involve both formal and informal caregivers, boost provider productivity, and offer a personalised and integrated experience. A single business model and a virtual data infrastructure connect the actors in a value chain, which include suppliers, customers, platforms, and service providers. Improved and more efficient experiences for stakeholders and customers are the outcome of the seamless collection, handling, and sharing of data that support this relationship. The term "ecosystem" refers to this networked and effective structure that is purposefully created to handle significant problems or inefficiencies in the system.

Usually, avoiding and successfully treating chronic illnesses are the main priorities. But when these objectives change, as we have seen, productivity in the healthcare sector falls behind other service sectors [2]. New technologies offer real-time metric-based care, lower costs among collaborated stakeholders, and the required treatment that is provided near to or at home [3]. Stakeholders must now shift to an ecosystem model of care, which is made feasible by five significant industrial forces driving technological innovation:

DOI: 10.4324/9781003475309-3

- Persistent inefficiencies in the industry contribute to challenges related to affordability, outcomes, and quality, resulting in a subpar consumer experience. These inefficiencies, existing over an extended period, create an opportune environment for innovation to yield substantial returns.
- Notable investments have been made in healthcare technology; from 2014 to 2018, over 580 US deals valued at greater or equal to $10 million each made a total of over $83 billion in overall earning value. Three main areas that received disproportionate attention from these investments were: innovative care models, data and analytics, and patient engagement.
- Tech behemoths are fiercely vying for supremacy in the cloud technology, investing substantial sums in research and development to enhance their platforms. The objective is to deliver user-friendly services catering to a broad customer spectrum and diverse applications, including areas like predictive analytics, in a bid to propel innovation. This intense competition amounts to a trillion-dollar battle, emphasising the strategic significance of capturing market share and sustaining consumer engagement. Furthermore, strategic partnerships, whether formed among pharmacy providers, health systems, or collaborations with technology firms, signify a growing trend towards increased integration. This movement also highlights a mounting emphasis on patient privacy concerns. Both traditional healthcare incumbents and emerging players can seize this innovation as a substantial opportunity to expand their market presence, all the while enhancing the cost-effectiveness and quality of healthcare delivery.
- In order to increase their data and analytics capabilities, payers, providers, healthcare services, and technology companies made strategic acquisitions totalling almost $40 billion in the healthcare sector between 2014 and 2018. This signalled the start of an important trend in which industry leaders invested heavily to develop their ecosystems as a whole.

In 2021, health tech fundraising broke previous records, totalling $29.1 billion over 729 deals, with an average confirmed deal of $39.9 million (as depicted in Figure 2.1) [4]. An investor has highlighted the current period as one of the most exciting times for health tech investments, and the market prognosis points to a continuous growth trajectory in the following years. In 2021, R&D catalyst ($5.8 billion), on-demand healthcare ($4.5 billion), and illness therapy ($4.5 billion) were the three main investment topics.

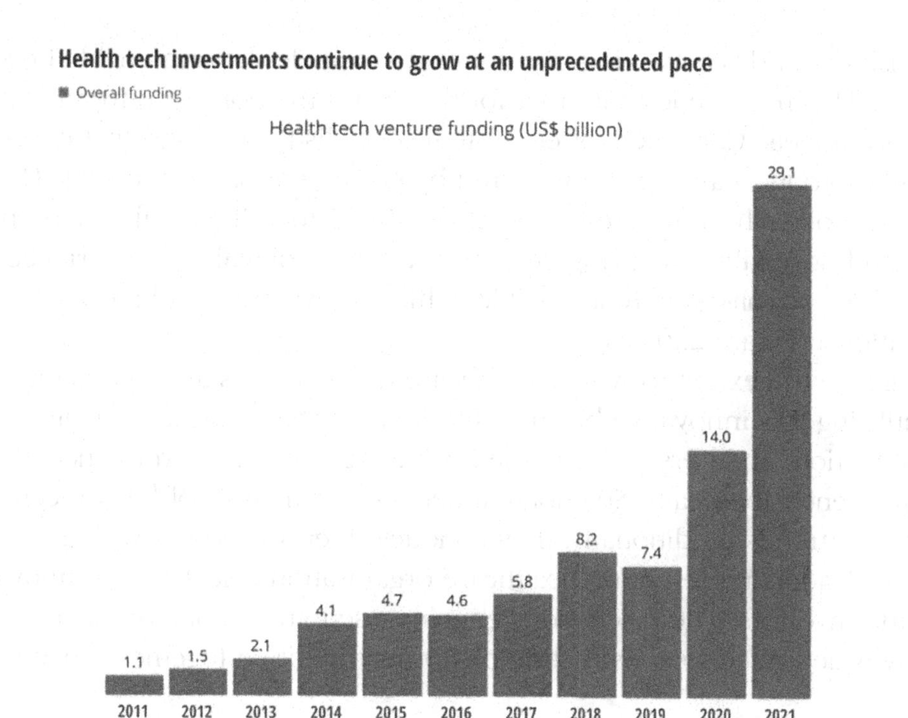

Figure 2.1 Chart showing the growth of healthcare domain investments [4].

However, some health finance databases discovered that, in addition to infrastructure and skill pools, modifications in organisational models and a greater emphasis on deliver-to-consumer and even business-to-business may have the greatest impact on the sector in the long run. Investments in platform businesses have already begun to come from venture capitalists, healthcare organisations, and private equity firms. For example, VEDA, an AI-based data management platform that leverages intelligent working to speed up administrative procedures in healthcare, has received $45 million.

Four of the nine healthcare businesses that filed for public offering in 2021 identify as platforms. However, not every company claiming to be a platform will succeed. The markets will value them differently depending on whether they are genuine platforms or not. Numerous healthcare companies have demonstrated that they should expand or diversify in response to opportunities (such as growing their market share) or risks (e.g., changing reimbursement models). However, due to limitations with resources, skill, and organisational structure, mergers and acquisitions might not always be the optimal course of action. Recent technological developments have made it possible for healthcare organisations to accomplish comparable goals (expansion and diversity) via a more recent channel: platform-enabled ecosystems. This has opened the door for the transformation of conventional business models. Conveners of ecosystems establish a platform that combines the products and services offered by ecosystem participants [5]. These business models have rewired value chains in sectors like retail, entertainment, and hospitality by giving customers a better, digitalised experience and making suppliers more accessible, which draws more and more users to the platform (Figure 2.2).

To gauge the extent to which healthcare organisations are embracing and utilising this innovative business model for enhancing care quality, cost reduction, and personalising the healthcare consumer experience, the Deloitte Center for Health Solutions undertook an analysis of health tech investment trends. Additionally, they conducted ten interviews with a varied group of leaders representing healthcare organisations, health tech innovators, and investors. The discussions encompassed an exploration of ongoing initiatives dedicated to constructing platforms that bring together ecosystems,

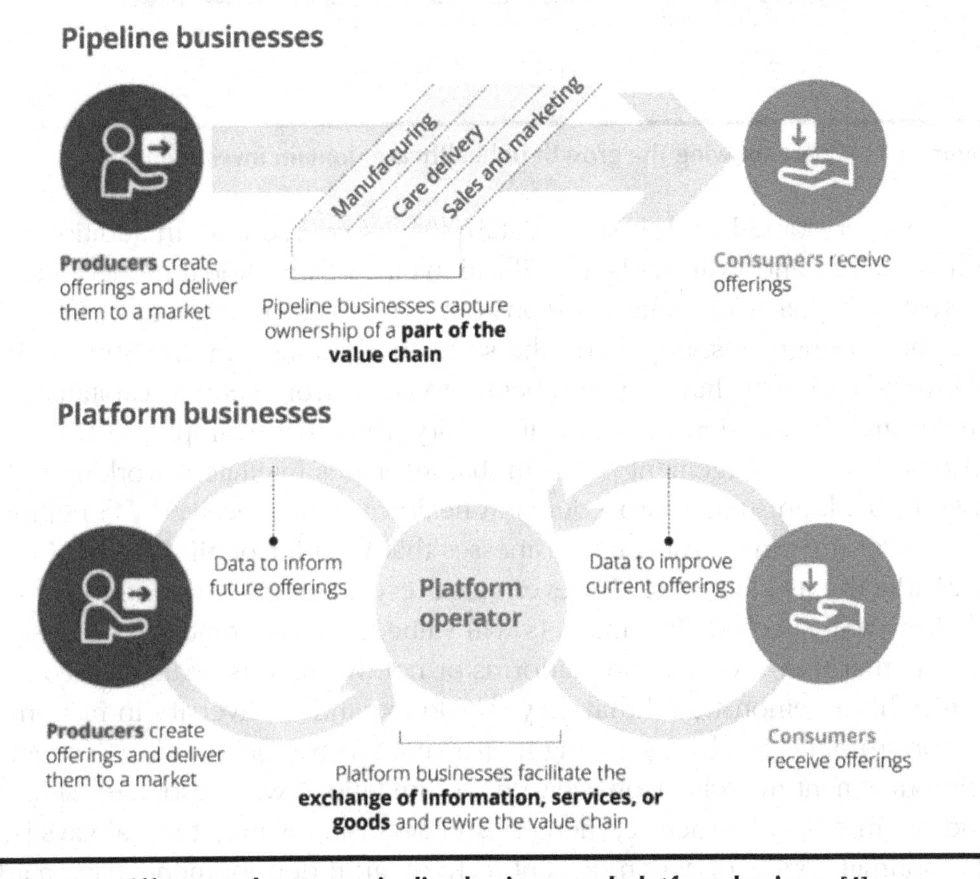
Pipeline businesses
Manufacturing
Care delivery
Sales and marketing
Producers create offerings and deliver them to a market
Pipeline businesses capture ownership of a part of the value chain
Consumers receive offerings
Platform businesses
Data to inform future offerings
Platform operator
Data to improve current offerings
Producers create offerings and deliver them to a market
Platform businesses facilitate the exchange of information, services, or goods and rewire the value chain
Consumers receive offerings

Figure 2.2 Differences between pipeline business and platform business [4].

the initial value derived from these efforts, and insights gained from these experiences.

2.2 Principles for Enabling Effective Ecosystems

Interviewees' responses to the idea of platform-enabled ecosystems in healthcare varied widely. Some were dubious about the idea of platforming healthcare, or even certain subsegments of it, and about ecosystems as a potential future economic model [6]. Rather, they favoured concentrating on developing technological remedies to tackle specific issues in the treatment process or establishing an edge pipeline enterprise as a substitute source of income. Some were still figuring out how to create an ecosystem, but they were interested in doing so. One of the respondents talked about creating community health-focused localised ecosystems, but encountered two obstacles: developing the technical foundation and persuading clinical and administrative leadership. Many of the interviewees, nevertheless, were enthusiastic about the idea. The following lists five guidelines that can support a productive, long-lasting platform-based ecosystem (Figure 2.3).

Platform principles in action
01
Accessibility of underutilized assets
Platform businesses facilitate large value-chain transformations and unlock value.
02
Delegation to ecosystem
By delegating nonstrategic assets to the ecosystem, platforms can scale rapidly.
03
Modularized components
Modularity and standardization make it possible for partners in the ecosystem to plug and play into the platform.
04
Focus on consumer experience
Enables the simplification and creation of an intuitive and convenient experience for users to drive adoption of the platform.
05
Positive network effects
A focus on consumer experience aggregates more consumers to the platform, which brings more producers, unlocking virtuous cyclical value.
Source: Ben Thompson, "Stratchery," accessed February 1, 2022.
Deloitte Insights | deloitte.com/insights

Figure 2.3 Principles for enabling effective ecosystems [4].

Availability of Untapped Resources: It is commonly accepted that the healthcare sector has valuable assets that are frequently not given enough attention. These assets include clinical knowledge, availability of healthcare providers, as well as diverse data sources like fragmented clinical and claims data and factors that influence health drivers (known as social determinants of health). Those interviewed highlighted the significant impact of creating platforms that enable ecosystems, allowing organisations to access and maximise the inherent value of these assets that have been underutilised [7]. For example, telemedicine platforms have successfully adopted underutilised provider availability, enabling clinicians to communicate with patients at any time and from any location, especially in a safer setting during the pandemic. The idea that data are underutilised assets was also covered in detail, with real-world examples of businesses realising this potential and creating a marketplace to connect data suppliers and buyers.

Empowering the Ecosystem through Delegation: There is a myriad of advantages in uniting participants through this innovative business model. For instance, the expanding influence of accountable care organisations is directly tied to the root causes of health, impacting the risk profile of hospitals. While hospitals may not directly address needs like housing, groceries, or job training, strategic partnerships with community-based organisations, facilitated by technology, streamline patient connections with these vital services. Furthermore, health plans are increasingly outsourcing disease management to virtual-first providers, eliminating the need for the health plan to independently construct such solutions, thereby facilitating rapid scalability for improved outcomes. In this dynamic landscape, organisations are leaning towards collaborative relationships with suppliers facilitated by networked platforms rather than pursuing outright ownership, offering a more efficient avenue for realising their strategic objectives.

Component Modularity: Platform-centric businesses may construct interoperable, standardised platforms that facilitate the integration of additional capabilities or specialised solutions, ultimately simplifying and refining the user experience. These modular components may be developed internally or by external partners. According to many interviewees, the adoption of modular platforms can eliminate redundancies, such as the development of multiple diabetes management apps. The concept of allowing partners or third parties to seamlessly integrate and utilise the platform for enhancing existing offerings and solutions, often described as "plug and play," was a common theme in the discussions. As an illustration,

HealthVerity's platform, focused on identity, privacy, governance, and exchange, empowers partners to amalgamate diverse datasets securely, enhancing record-matching accuracy. This platform was strategically designed to accommodate various datasets, standardise intricate processes, and capitalise on the network effect, thereby expanding its user base through both data suppliers and buyers.

Prioritise the Patient Journey: The responders stressed the significant achievement of enhancing and simplifying the consumer experience in platform-based ecosystems. One participant highlighted, "Ultimately, the goal is to enhance consumer convenience." Platforms not only create a digital interface for care delivery but also optimise the omnichannel care experience for patients. A recent study showed that a majority of health system digital executives view digital capabilities as crucial for transforming their interactions with consumers.

Synergistic Impact of Networks: Distinguishing themselves from pipeline businesses, platform businesses rely on virtuous network effects for competitive advantage. These platforms foster growth by incentivising ecosystem partners to join and address existing gaps. Prominent consumer technology firms often leverage demand-side strong network effects, where an expanding user base enhances the value they deliver, leading to improved offerings that are either superior or more cost-effective. Access to more user behaviour data enables businesses to refine their products or services, creating a cycle where an enhanced offering attracts additional consumers and partners [8]. Additionally, producers within the platform can collaborate; for instance, on a care-management platform, virtual-first diabetes companies can engage with virtual-first behavioural health companies.

2.3 Factors to Contemplate in Platform-Enabled Ecosystems

Healthcare entities seeking to strategically position themselves for the Future of Health™ and address ongoing influences such as COVID-19, value-based care, data liquidity, consumer expectations, digital transformation, and virtual health, should explore the concept of assembling a platform or becoming part of an ecosystem. In formulating their approach to leveraging platforms and ecosystems, organisations should take into account the following essential considerations:

- Recognise opportunities to expand the value chain (e.g., enhance outcomes, improve customer experience, and lower costs) without resorting to mergers and acquisitions (M&A): With the increasing responsibilities placed on healthcare organisations, leaders should assess areas where collaboration within an ecosystem and utilising a platform might prove advantageous. Entrusting certain aspects to partners can serve as a cost-effective alternative to extensive vertical integration, often yielding results that are equally, if not more, successful.
- Spot opportunities to transform threats into advantages: In the face of business threats, introducing a platform can effectively neutralise those challenges. For instance, many health systems are assuming additional risk and actively seeking avenues to tackle the fundamental determinants of health. Through establishing an ecosystem in collaboration with a network of community-based organisations, health systems can more effectively address the underlying factors influencing their patients' health.
- Establish alliances within the ecosystem: Identifying suitable partners is a crucial aspect of building an ecosystem. Each entity brings unique skills and offerings, necessitating alignment with the overarching goals and purpose of the ecosystem. Additionally, each participant must determine their preferred roles within the ecosystem. Will they act as a convener, taking charge of platform creation, operation, and interaction orchestration? Alternatively, would a participant role, involving the provision of goods and services on the platform, be a more suitable fit?

In the past, pipeline business models have been the norm for life sciences and healthcare organisations. These businesses competed on the basis of price, quality, and market share, with M&A serving as the main growth path. However, it is becoming unfeasible for a single healthcare organisation to provide the full range of care, from clinical services to addressing fundamental health drivers, spanning well-being to long-term care, given the expanding breadth of healthcare products and services [9]. Using a platform business model allows for similar results with lower capital expenditures, leverages the sharing of goods, services, or information among ecosystem participants, and competes on the basis of customer experience and network effects.

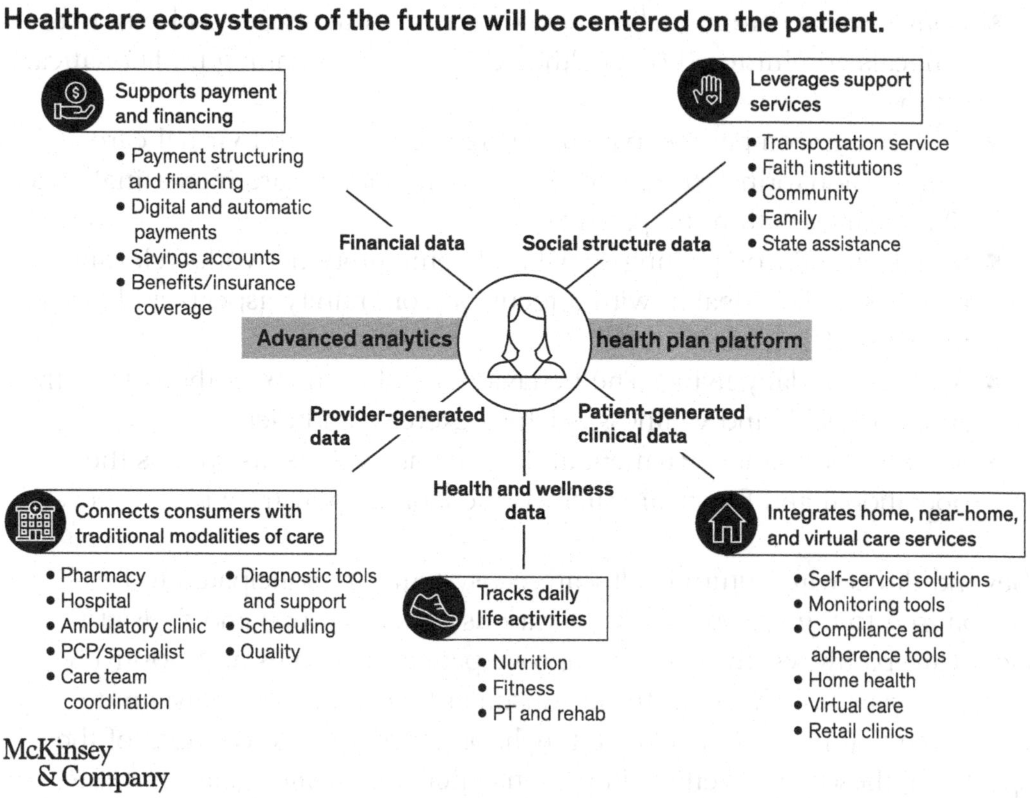

Figure 2.4 Patient-centred healthcare ecosystems [10].

2.4 Potential Prospects for Healthcare Ecosystems

All sectors have developed ecosystems, and both incumbents and technology disruptors benefit from them. Disney is one such. Because of its strong ecology, every part can support and strengthen the others. Disney released its first motion picture in 1937, its first TV show in 1954, and its streaming service Disney+ in 2019. Characters are reinforced at its theme parks, like Disney World, giving families and kids fun, in-person encounters. A self-reinforcing experience inside the ecosystem is created when those youngsters ask for Disney games, Disney toys, and Disney attire. This is made possible by the control over a limited resource, content, as well as the underlying data and analytics to best distribute it.

In the future, healthcare ecosystems, akin to other ecosystems, will revolve around the consumer, specifically the patient. The capabilities and services constituting the healthcare ecosystems of the future (Figure 2.4) will encompass, among other things:

- Conventional care modalities: Encompassing direct care and pharmaceuticals administered by healthcare providers, spanning traditional care settings.
- Self-care: Encompassing patient engagement, self- and virtual care, remote monitoring, and the delivery of traditional care increasingly feasible near or within the patient's home.
- Social care: Encompassing social and community networks relevant to a patient's holistic health, with a focus on community aspects addressing unmet social needs.
- Activities of daily living: The behaviours and routines of the patient that support health and wellness, such as exercise and diet.
- Support for finance: Payment and financing options, as well as the operational and financial framework enabling industry-wide care events.

Beyond direct care, future healthcare ecosystems are anticipated to be customised to the various needs of various patient groups and their efficient care pathways. In order to improve patient outcomes and positively influence patient behaviour, these consumer-focused ecosystems aim to expand the number of healthcare touchpoints [11]. On the one end of the spectrum, these ecosystems will prioritise personal health goals and cater to the needs of healthy patients who face fewer ongoing medical challenges. This describes an ecosystem that is primarily digital and that uses wearable technology to enable highly personalised use of patient data and insights. In this scenario, the percentage of total touchpoints that involve traditional care modalities is anticipated to be relatively low.

At the other end of the spectrum, healthcare ecosystems are about to come into their own to meet the needs of patients who are managing several complex chronic illnesses. More specifically, for those who fall into the dual-eligible Medicare and Medicaid categories, efficient provider coordination and in-person and virtual service delivery (at home or nearby) are critical for a smooth end-to-end process. It is anticipated that technological components of these ecosystems will be used to support the comprehensive care team and improve the in-person experience [12, 13]. Informal caregivers—adult children of elderly patients, for example—may play an increasingly important and technologically enabled role in this team. Numerous healthcare startups are currently investigating and testing this focused model.

Ecosystems are structured with three fundamental layers: infrastructure, intelligence, and engagement. The foundational layer, referred to as infrastructure, involves efficient processes for capturing, curating, managing, storing, and ensuring interoperability of data. These elements collectively establish a standardised dataset that serves as the underpinning for the ecosystem's functioning. Positioned above the infrastructure layer is the intelligence layer, tasked with converting data elements into insights that are both understandable and actionable. Ultimately, to actualise an ecosystem, a robust engagement layer is imperative. This layer, facilitated by the infrastructure and intelligence layers, adeptly shapes a comprehensive experience for suppliers offering services and solutions to patients. The constituents of these layers can be developed, procured, collaboratively pursued, or outsourced by ecosystem curators and participants.

As they invest billions of dollars in R&D to create cross-industry capabilities, digital giants will shape the evolution of the healthcare ecosystem [11]. The only thing left to ask is what function it will serve. At every level, they impart at least the fundamental knowledge. In this future state of affairs, the leading healthcare companies will curate ecosystems and expand upon the capacities of the big digital players by providing industry-specific services that improve their engagement, intelligence, and infrastructure that spans horizontal and cross-industry boundaries. For instance, tech behemoths are fighting for supremacy in the cloud technology and developing infrastructure that can give healthcare providers access to enormous amounts of processing power.

Additionally, these digital behemoths will be able to expand this infrastructure as data becomes more liquid, giving healthcare stakeholders a solid base upon which to build in order to capture value through healthcare ecosystems and gain market share versus more established rivals. Healthcare stakeholders that collaborate with these tech behemoths will also need to handle the major risks that come with these agreements, such as those pertaining to internet protocol (IP) regulation, privacy, and confidentiality.

It's possible for the biggest digital companies to create and manage their own ecosystems. They might enter the market by going direct to customers or by forming alliances with a small number of well-established companies. Healthcare profit pools in this environment will probably be upset as these big digital companies disintermediate the patient-provider connections of current healthcare incumbents or provide authority to incumbents who are collaborating with them to increase their share. Large technology companies would probably need to see a number of regulatory changes (such as boosting data interoperability) and have a reasonable amount of confidence in the economics of upending the industry rather than encouraging innovation within it, which usually carries less risk, in order for this scenario to

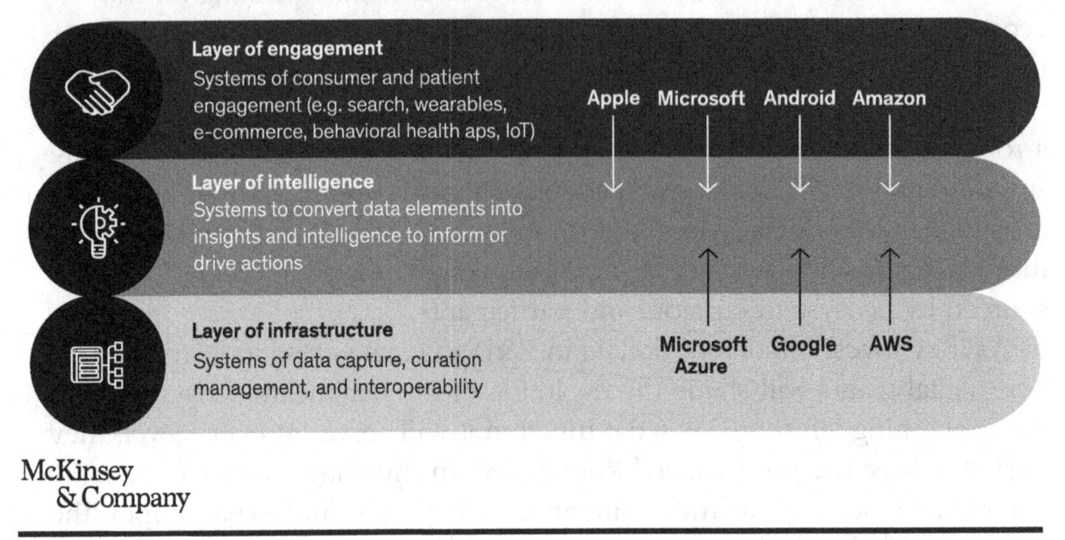

Figure 2.5 Investments of technology giants across healthcare ecosystems [10].

materialise. As discussed above, Figure 2.5 shows the investments of technology giants investing across the layers of ecosystems from the healthcare domain.

While some are leading the way, many of the major players in the healthcare industry have not yet publicly disclosed their ecosystem strategy. Additionally, they lack the staff, infrastructure, and technological capabilities required to fully realise the benefits of curating or participating in patient-centred, multi-site ecosystems. It is up to stakeholders to carefully consider whether they wish to curate or participate in the ecosystems they influence. Stakeholders must guarantee a notable improvement in outcomes for a particular patient population in order to curate an ecosystem [10, 14].

To establish a unique ecosystem, this strategy will need to be transparent about the industry-neutral services it uses and how it enhances them with capabilities unique to the healthcare sector. If other stakeholders wish to offer point solutions, they must make sure that their value proposition is adaptable to different ecosystems and competitively different.

The majority of participants in healthcare technology and services will probably operate as (one or many) point solutions that plug into and perform essential tasks inside changing healthcare ecosystems. As a result, these participants ought to increase the variety of circumstances and environments in which they may use their skills. Businesses frequently connect solutions inside or between subject domains that may extend beyond an ecosystem's surface layer. Technical adaptability and APIs that communicate with other ecosystem members are necessary for this action, particularly in the intelligence and core infrastructure layers. Technology companies and healthcare providers could also develop modular solutions that make it easier to expand an ecosystem's capabilities.

In the end, some tech firms and healthcare providers might be able to assemble effective sub-ecosystems. This is less common than implementing point solutions, which concentrate on a particular use case (like payments) or population (like diabetics). A patient engagement technology that targets a particular disease, for instance, might presume that it can directly create a setting for patients at a particular degree of patient involvement. In order to implement this strategy, participants would need to provide the necessary intelligence and data infrastructure in addition to curating a full or nearly full spectrum of physical and digital services geared to diabetes patients.

References

1. Micalo S. Healthtech innovation: how entrepreneurs can define and build the value of their new products, (1st ed.) 2022. Productivity Press. https://doi.org/10.4324/b23147
2. The untapped potential of ecosystems in health care | BCG n.d. https://www.bcg.com/publications/2021/five-principles-of-highly-successful-health-care-ecosystems (accessed January 15, 2024).
3. Page W, Garbuio M, Wilden R. The role of incubators and accelerators in healthcare innovation. *Healthc Entrep* 2018:87–107. https://doi.org/10.4324/9781315157993-5.
4. Transformed health care ecosystems | Deloitte insights n.d. https://www2.deloitte.com/us/en/insights/industry/health-care/transformed-health-care-ecosystems.html (accessed January 15, 2024).
5. Cosimato S, Di Paola N, Vona R. Digital social innovation: how healthcare ecosystems face Covid-19 challenges. *Technol Anal Strateg Manag* 2022. https://doi.org/10.1080/09537325.2022.2111117.
6. Naughton B, Dopson S, Iakovleva T. Responsible impact and the reinforcement of responsible innovation in the public sector ecosystem: cases of digital health innovation. *J Responsible Innov* 2023;10. https://doi.org/10.1080/23299460.2023.2211870.
7. Alzate Montoya MA, Montenegro Martinez G, Londoño Pelaez C, Cardona Arango D. Innovation ecosystems in health: countries and theoretical models used. *F1000Research* 2022;11:1458. https://doi.org/10.12688/f1000research.125854.1.
8. Klinedinst J.(Ed.) The handbook of continuing professional development for the health IT professional, (2nd ed.). 2021. .Productivity Press. https://doi.org/10.4324/9780429398377

9. Thapa RK, Iakovleva T. Responsible innovation in venture creation and firm development: the case of digital innovation in healthcare and welfare services. *J Responsible Innov* 2023;10. https://doi.org/10.1080/23299460.2023.2170624.
10. The next wave of healthcare innovation | McKinsey n.d. https://www.mckinsey.com/industries/healthcare/our-insights/the-next-wave-of-healthcare-innovation-the-evolution-of-ecosystems#section-header-1 (accessed January 15, 2024).
11. Kolasa K. The digital transformation of the healthcare system : healthcare 5.0, (1st ed.). 2023. Routledge. https://doi.org/10.4324/b23291
12. Foglia E, Garagiola E, Bellavia D, Rossetto F, Baglio F. Digital technology and COVID-19 pandemic: feasibility and acceptance of an innovative telemedicine platform. *Technovation* 2024;130:102941. https://doi.org/10.1016/J.TECHNOVATION.2023.102941.
13. Cozzolino A, Geiger S. Ecosystem disruption and regulatory positioning: entry strategies of digital health startup orchestrators and complementors. *Res Policy* 2024;53:104913. https://doi.org/10.1016/J.RESPOL.2023.104913.
14. Top cities for Healthtech startups in 2023 n.d. https://www.startupblink.com/blog/top-cities-for-healthtech-startups/ (accessed January 15, 2024).

CONCEPTION

II

Chapter 3

Idea Generation and Need Analysis for Healthcare Products

3.1 User Integration

Every invention begins with an idea that inevitably spreads [1]. The ideation stage represents the artistic process of generating, developing, and evaluating new ideas. As such, ideation provides the catalyst and foundation for creating prototypes and breaking new ground in creative solutions. The ideation process is divided into the following four stages. Identifying current obstacles and client requirements is the first step. The second stage then involves coming up with new concepts. The evaluation and selection of workable concepts are the main focuses of the third phase. Prototype development is the final step in the process. Approaches and instruments for methodical brainstorming take on increased importance, especially when considering cross-sectoral patient care. The need for more innovation stems from the difficulties that arise when coordinating across various sectors. These difficulties include decreased patient adherence and increasing performance and cost pressures, which are best illustrated by the growing number of elderly people and the significantly higher number of patients who are chronically ill with multimorbidity.

Securing access to users and consumers, in particular patients and healthcare professionals, is the main difficulty facing the healthcare sector. Addressing the gap between consumers and businesses is essential. Diverse

DOI: 10.4324/9781003475309-5

techniques and resources are essential, especially during the ideation stage of the innovation process, as they enable alignment with the needs of pertinent target audiences while also satisfying the needs of businesses looking to launch innovations in the health sector [2]. This two-pronged approach lays the groundwork for supporting both innovative projects and long-term developments that raise the standard of patient care.

Growing past the advantages and real requirements of the target groups is a major issue facing the medical business. Low customer or user engagement in the ideation process is often a contributing factor, along with inadequate identification of the necessary criteria. These days, physicians—chief physicians in particular—are frequently only partially considered as potential innovators, and most businesses lack a structured procedure for including pertinent and active stakeholders in the ideation, assessment, and development process.

Several conventional methods of user involvement face constraints because of the unique features of the healthcare industry. Apart from the non-medical health workers (such as nurses, therapists, etc.) who function as professional end users, patients are another group that the healthcare system frequently ignores. Their inadequate understanding of medical technology and regulatory laws limit their potential as innovators, even though they play a vital part in the ideation process [3]. Thus, focused and intentional participation becomes essential. Prior to the purposeful incorporation of users, businesses need to answer critical questions: when and how should certain customers/users be actively involved in the innovation process? (see Figure 3.1).

Therefore, a variety of elements influence the particular user integration technique selection. For instance, based on the chosen ideation phase and the intended integration goal, distinct user/customer types should be included. Some instances are:

1. Generation and assessment of ideas: Working with trendsetting, creative users who have the necessary specialised knowledge is necessary to identify user requirements and formulate sustainability-driven ideas and concepts (lead users).
2. Field testing and prototyping: To validate and iteratively improve prototypes, users must be included in real-world usage scenarios with experts and/or other representative users from all relevant fields.
3. Collaboration: Working together with pilot customers that are focused on sustainability (such as important clients like clinics) with the same goal of presenting new and creative solutions.

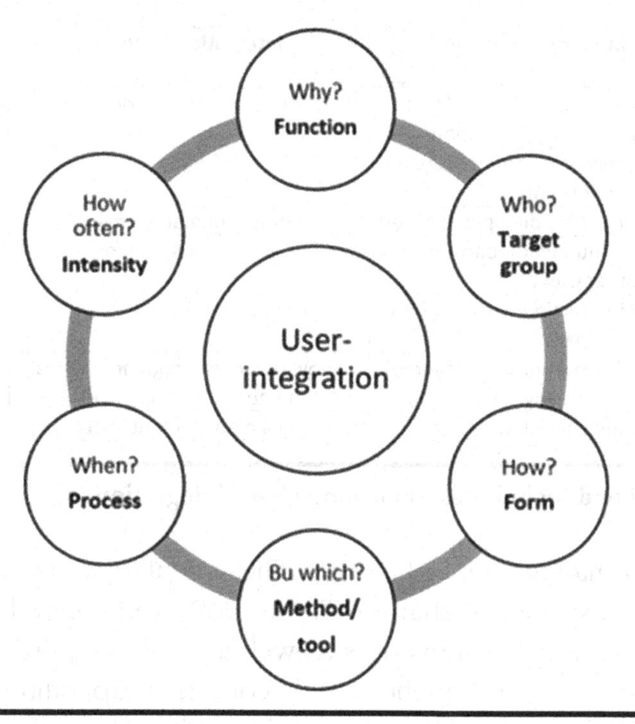

Figure 3.1 User integration [4].

The target group determines the manner of integration, or the "how" question. Both traditional and practical methods can be used to assess the level of integration (see Figure 3.2).

A diverse array of methodologies and tools is at the disposal of companies to systematically engage patients and other pertinent target groups in their internal ideation, thereby fostering the development and implementation of more user-centric and sustainable innovations [5]. The prevailing approaches typically encompass innovation workshops and digital idea management tools, which represent the most commonly utilised variants:

- Creativity workshops: Participants can include relevant users like doctors and other healthcare professionals, as well as internal and external experts and even patients. These meetings seek to thoroughly examine current issues and needs with a multidisciplinary approach, then contrast them with the demands and ideas of manufacturers (companies). Innovation workshops may also evaluate concepts, where finished idea concepts or prototypes are shared and debated as a group (feasibility study), with the possibility of additional improvement. Innovation workshops provide a wide range of features and structure options. Businesses might choose to self-sponsor these kinds of seminars or enlist the help of other innovation management specialists.

The integrated user, function	Integration method	Amount of Integration
User as a passive observation and test object without direct interaction	Field tests and observations	low
User as an externally determined dialogue partner survey	Survey	
User as a self-determined dialogue partner	Complaint management	
User as an equal interaction partner of other users	Communities und fora	
User as an equal interaction partner of the company	(Innovation-) Workshops	
User as an equal (innovation) employee of the company	Innovation circle (regular informal meetings for exchange)	
User as an independent innovator	Idea management tools	high

Figure 3.2 Level and techniques of customer/user integration [4].

- Digital idea management platforms: These enable constant information sharing and experience sharing. These tools are intended to inspire staff creativity and inventiveness as well as those of pertinent users and consumers. Typically, these tools comprise capabilities for recording, regulating, conveying, and assessing ideas. These technologies can also be tailored to meet the unique requirements of the business. Idea management solutions are not just for gathering and ranking fresh ideas or suggestions for improvements that make sense. In order to realise sustainable innovation potential and contribute to the company's long-term success, it also entails strengthening internal innovation processes.

Idea creation, also known as ideation, is the initial stage of the product development process that involves gathering data and original thinking to produce concepts for new products [6]. A strong foundation is laid during the ideation stage, which steers the new product concept in the direction of success. However, coming up with great ideas is difficult, and putting them into practice is much harder. It might be difficult for even the most creative teams to achieve that elusive "lightbulb moment." But, it is possible to produce something truly unique by using the right methods and strategies. During the ideation phase, an entrepreneur receives the following benefits:

- Tackle fundamental inquiries: Noteworthy challenges within the new product development process emerge during the ideation phase. Essential questions, including the distinctiveness of this new product idea, the identification of the target customer, and the efficacy of the idea in resolving customer problems, demand attention from the team.

Resolving these queries fosters a deeper understanding of customer needs and the overarching objectives in play.

- Utilise the creativity within the team: Product ideation is a collaborative effort. The collective creative contributions of team members prove invaluable in generating a new product idea. Ideally, an adept team comprising approximately two to eight subject matter experts (SMEs) should partake in an ideation session. For instance, a standard session might incorporate a product owner, product manager, a designer, and a software engineer. This composition facilitates the exchange, scrutiny, and synthesis of a diverse array of perspectives, cultivating a reservoir of promising ideas.
- Prioritise the customer: Product ideation extends beyond conceiving the optimal product concept; it involves selecting one that aligns with your company's strategic objectives, and, most significantly, addresses the needs of your potential customer. The team must consistently place the customer at the forefront of their considerations while exploring a new idea [7].

3.2 Origins and Advantages of Concept Creation

Acquiring data from many sources is essential for pinpointing (and eventually resolving) your clients' problems. Businesses frequently use many sources at once, some of which will be more beneficial than others. Here are few instances:

- Leverage industry expertise for automation opportunities: Tap into the insights of team members or SMEs with backgrounds in different industries (e.g., hospitality, retail, healthcare). They might pinpoint areas within those sectors where internal processes or customer products/services could be automated or enhanced. Subsequently, validate these observations through market research. This approach not only provides understanding of evolving customer behaviour and emerging trends but also ensures that decisions are grounded in data. When an idea advances for approval, the readily available data can substantiate its feasibility and potential return on investment (ROI).
- Improve on what already exists: Sometimes it's not necessary to start from scratch with a solution; instead, concentrate on making it better. Start by analysing the competitors in great detail. Are there any areas where their offerings fall short? If not, listen to what their consumers

have to say and find out where your company might improve. Even in the unlikely event that no new product concept materialises, you could learn how to improve your present offerings.

- Solve organisational problems by using solution marketing. Determine the real-world problems that your teams are facing. Have they looked everywhere for a solution to their issue and been unsuccessful in their search? Examine the potential for developing a device or appliance to tackle and overcome these difficulties.
- Make use of internal teams' and stakeholders' knowledge: Everyday customer-facing personnel, such as R&D, sales, marketing, or engineering teams, have important insights. As a result, it is crucial that the company fosters an innovative culture and provides chances for team members to share ideas. This strategy encourages staff members to candidly address issues or potential fixes that might become brand-new product concepts.
- Respond to the needs of current clients: Customer input may spark the creation of new items as well as point out areas for product improvement. Consider the remarks or concerns voiced by your existing customers. Have they put out a concept for a product that might address a gap in the market? Some suggestions could point your team in the correct direction, even though not all of them are worth pursuing.

Despite its seeming contradiction, the idea of "structured creativity" works well for ideation of new products. A planned idea-generation session may provide better outcomes than an unstructured brainstorming session for a number of reasons.

- Structure enhances session productivity. Efficiency is compromised when ideas are shouted out and jotted down on a whiteboard. Employing a framework enables the team to share suggestions more effectively and maintain focus on the task at hand.
- Session structures can spark creative ideas. Typically, ideation frameworks include headers or prompts that encourage answers to pertinent questions. Using a framework-based organised approach helps individuals see things from several angles, which encourages creativity.
- Setting up frameworks early on establishes the right atmosphere for the complete undertaking. Whether getting ready for the launch of the finished product or concept validation, the team may methodically evaluate an idea and create an implementation strategy with a clearly defined framework.

3.3 Common Structures for Conceptualising New Products

Upon defining the problem, concluding research, and consolidating findings, the subsequent step is to plan an ideation session [8]. This planning, inclusive of tasks such as dispatching calendar invitations and selecting collaboration tools, necessitates choosing a framework that empowers the team to shape a new product idea. It is noteworthy that such tools find applicability across diverse tasks, regardless of the organisation's size.

SCAMPER: SCAMPER is a logical thinking method for ideation that stresses a fresh viewpoint on current ideas with a business-specific methodology. SCAMPER, which consists of seven different methods, is an acronym for "substitute, combine, adapt, modify, put to another use, remove, and reverse." These phrases encourage the re-examination of all aspects of an issue, leading to the discovery of the best possible solutions (see Figure 3.3).

SWOT Analysis: Strengths, Weaknesses, Opportunities, and Threats, or SWOT, is an acronym that is useful for brainstorming as well as concept screening. A SWOT analysis starts with choosing a topic or goal, such creating a new tool for social media management. Listing the goal's attributes comes next once it has been determined. Make a list of the subject's advantages and disadvantages before delving into its underlying strengths and shortcomings. Concurrently, list its Opportunities and Threats for an analysis of outside influences on its performance. To find their weaknesses, it might also be helpful to perform a SWOT analysis on one or more competitors.

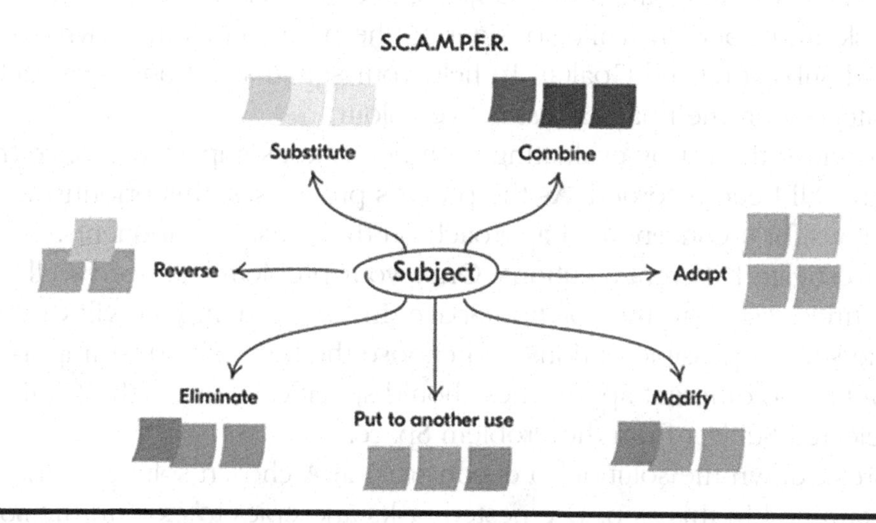

Figure 3.3 SCAMPER technique [9].

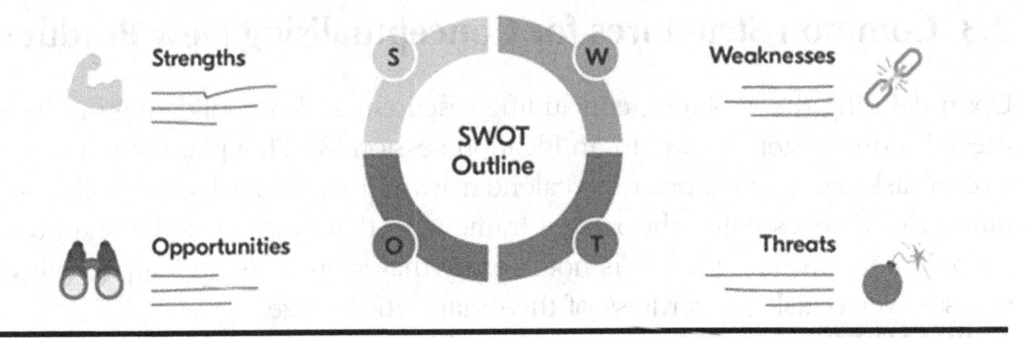

Figure 3.4 SWOT analysis template [9].

Every framework mentioned above has certain drawbacks. While both serve as helpful tools for brainstorming, none offers a direct route to the phases of selection or implementation of new product development. Furthermore, SWOT is more appropriate for the ideation screening stage because of its condensed form, which leaves less room for delving deeply into an issue (see Figure 3.4).

It is recommended to invite team members—designers, engineers, members of the product team (product manager, business analyst, etc.) and other stakeholders involved in the development process—to participate in the process. To get the best results, try inviting between two and eight people. The process of generating ideas might be conducted by following these stages.

- Start in the Problem Space to define the problem: List all possible subjects and provide a description of each. Users, organisations, roles, and tactics are examples of subjects. Record each and every Advantage, Risk, and Concern while considering the pertinent Domain Knowledge and Subject-related Goal(s). To help your squad stand out, give each category on the board a distinctive colour.
- Prioritise the list by evaluating each descriptor's importance once they have all been recorded. As the process progresses, this prioritising allows for a concentrated approach on the crucial components.
- Investigate Different Solutions: Once your problem has been well defined, go on to the Solution Space. In this section, you will examine the several possible options and choose the best one. Most importantly, the different approaches should specifically target the Goal of the selected Subject from the Problem Space.
- Break down the solution to deconstruct it: A chosen solution variation is examined in this step. Use nested tasks and epics to explain the solution in this last stage. To determine which Benefits, Risks, and Issues

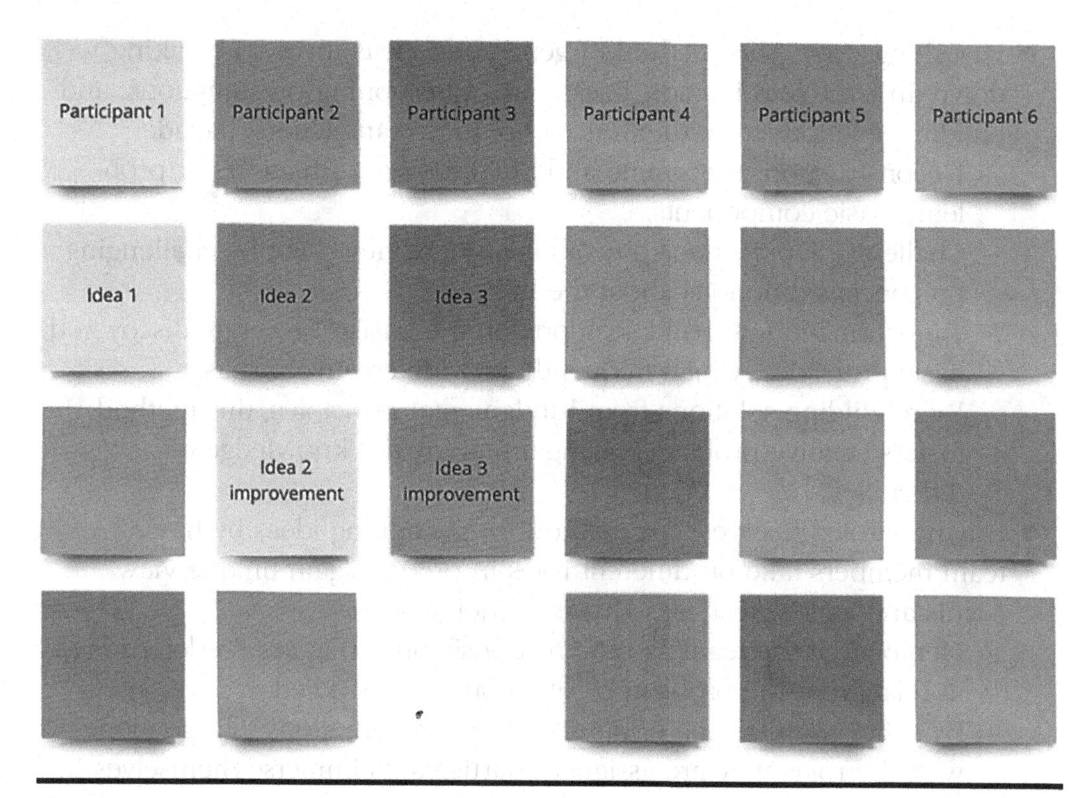

Figure 3.5 Brainstorming technique with cards [10].

belong on a task or epic, consult the Problem Space. Use the MoSCoW technique to assign jobs a proper priority. The MosCoW represents four categories of initiatives: must-have, should-have, could-have, and won't-have, or will not have right now.

- Brainstorming Methodology using Brainstorm Cards: Using this approach, each team member puts forward ideas on a separate card. Team members develop and expound upon the concepts that they get on their cards after shuffle and redistribute. A group discussion then takes place, exploring combined ideas and ranking ideas that show promise for further development. This method encourages teamwork and ensures that the brainstorming process benefits from a wide range of viewpoints (see Figure 3.5).
- Blue sky thinking: This method involves people thinking freely and imagining perfect situations without any limitations. The focus is on novel and unorthodox concepts, which encourage alternative thinking [8, 11]. The main objective is to come up with as many ideas as possible without immediately thinking about their viability. This fosters creativity by looking at options outside of the norm.

- First things first: This method of generating ideas involves breaking down an issue into its most basic parts, questioning presumptions, and building solutions from the bottom up. Important actions include:
  - Deconstruction: Determine and break down a situation's or problem's basic components.
  - Challenge Suppositions: Encourage a new viewpoint by challenging preconceived notions about the issue.
  - Transformation: Rebuild solutions on the basis of recently discovered basic principles, which frequently provide creative results.
  - By rebuilding solutions from fundamental principles, this method fosters creative problem-solving and a greater knowledge of difficulties.
- Playing a role: It serves as a method for generating ideas by having team members take on different roles in order to gain unique viewpoints and foster creativity. This is a brief synopsis:
  - Members of the team assume responsibilities that are predetermined in relation to the problem or situation being studied.
  - By fully assuming the opinions, ideas, and emotions that go along with the roles they are assigned, participants immerse themselves in the roles.
  - As a result of contact and discussion within the designated roles, fresh concepts, and insights frequently present unusual solutions.
  - After the role-playing exercise, participants evaluate their learnings and draw on creative concepts that sprang from the immersion process.

  By pushing participants to think beyond their typical viewpoints, role-playing encourages creativity and promotes a dynamic approach to problem-solving via embodied inquiry.
- Storyboarding: Another technique for coming up with ideas is to use a visual storytelling approach. Using this method, team members work together to create a visual storey that is step-by-step and allows for a thorough examination of the idea's evolution. This approach makes ideas come to life graphically and fosters teamwork as the group improves and refines the ideas that are shown. It is a creative process where concepts develop and take shape through a visual storey that has been collaboratively created.
- Mind mapping is a visual method for organising concepts around a main notion. Starting with the core idea, branches go to important concepts and then use keywords to branch deeper into sub-ideas. It

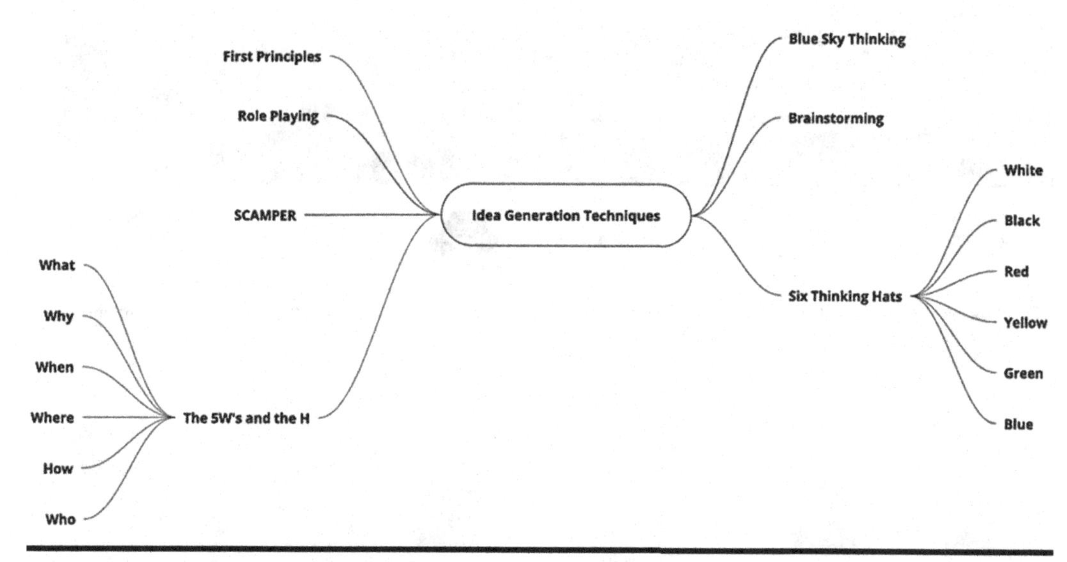

Figure 3.6 Mind mapping [10].

offers a non-linear framework with integrated graphic components that encourages creative exploration and connections between concepts. It enhances idea organising and brainstorming as a dynamic visualisation tool (see Figure 3.6).

- Opposite thinking: This method of generating ideas involves consciously considering the antithesis of typical ideas or behaviours in order to stimulate originality. It encourages people to deviate from standard thought patterns, which frequently results in the discovery of novel and unusual answers that would not be discovered by conventional thinking.
- Social listening: This strategy involves keeping an eye on internet forums in order to understand and evaluate discussions surrounding a company, a product, or a sector. Businesses monitor conversations, mentions, and trends on blogs, forums, social media sites, and other online forums in a methodical manner. They then analyse the sentiment of the chats to determine the prevailing views, inclinations, and patterns. Through this method, organisations may gain significant insights into the opinions, requirements, and worries of their customers, allowing them to customise their strategies to meet their expectations.
- Analogy thinking is a process for generating ideas that involves finding similarities between seemingly unrelated things in order to inspire original thought. Teams begin by drawing connections by finding commonalities between the current problem or idea and unrelated ones [12]. They then apply knowledge or solutions from the unrelated idea to the

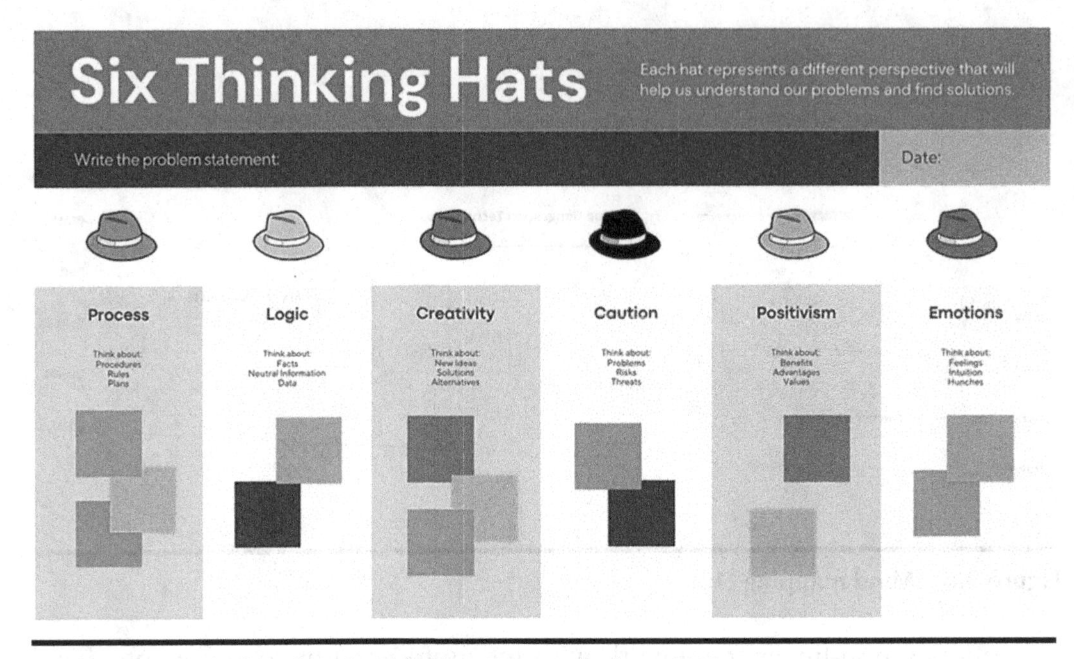

Figure 3.7 Template for Six Thinking Hats technique [10].

first problem. Analogies serve as a creative jumping-off point, connecting the known with the unknown to generate fresh viewpoints and concepts.

- Six hats for thought: Edward de Bono created the Six Thinking Hats, a systematic strategy for coming up with ideas and making decisions. Here's a quick rundown:
  - Six Perspectives: Give each participant a different "hat," each of which stands for a particular perspective: White for facts, Red for feelings, Black for critical judgement, Yellow for optimism, Green for creativity, and Blue for controlling the thought process.
  - Sequential Focus: By putting on one hat at a time, participants may think more systematically and intently. For instance, begin with White while acquiring information, then move to Red when considering emotions.
  - Role-Based Speculating: Every hat directs thought in a certain way, guaranteeing a thorough examination of a subject or choice from several perspectives.
  - Decreased Disputes: The Six Thinking Hats technique reduces conflicts, promotes positive communication, and aids in well-rounded decision-making by breaking thinking up into discrete roles (see Figure 3.7).

As the investigation into idea creation in business draws to a close, useful tools and strategies that foster creativity and problem-solving have been revealed. The evolution from foundational ideas to powerful analytics tools such as iSwarm demonstrates a broad toolbox that may empower teams. Innovation isn't just a fancy phrase in the corporate world; it's what propels advancement. It's about figuring out practical issues and changing with the times. So, equipped with useful knowledge and flexible instruments, let's design a future in which the innovative concepts result in real-world success.

References

1. Alcia L, Wan S. Increasing revenue through idea generation at university health network 2013;26:37–40. https://doi.org/10.1016/J.HCMF.2012.12.001.
2. Guilabert M, Sánchez-García A, Asencio A, Marrades F, García M, Mira JJ. Retos y estrategias para recuperar y dinamizar la atención primaria. Metodología DAFO (Debilidades, Amenazas, Fortalezas y Oportunidades)-CAME (Corregir, Afrontar, Mantener y Explotar) en un departamento de salud. *Atención Primaria* 2024;56:102809. https://doi.org/10.1016/J.APRIM.2023.102809.
3. Lee YSH, Grob R, Nembhard I, Shaller D, Schlesinger M. Leveraging patients' creative ideas for innovation in health care. *Milbank Q* 2023:1–38. https://doi.org/10.1111/1468-0009.12682.
4. Idea – Healthcare platform n.d. https://www.access-platform.eu/en/idea/ (accessed January 16, 2024).
5. Girotra K, Terwiesch C, Ulrich KT. Idea generation and the quality of the best idea. *Manage Sci* 2010;56:591–605.
6. Plsek P. Innovative thinking for the improvement of medical systems. *Ann Intern Med* 1999;131:438–44. https://doi.org/10.7326/0003-4819-131-6-199909210-00009.
7. Cañizares JCM, Rojas JC, Acuña A. Idea generation and integration method for inclusion and integration teamwork. *Front Educ* 2023;8:1009269. https://doi.org/10.3389/FEDUC.2023.1009269/BIBTEX.
8. Bani IA. Health needs assessment. *J Family Community Med* 2008;15:13. https://doi.org/10.4324/9780203068328-4.
9. Idea generation for new product development | Railsware Blog n.d. https://railsware.com/blog/idea-generation-for-new-product-development/ (accessed January 16, 2024).
10. Idea generation in business - 16 techniques | EPAM SolutionsHub n.d. https://solutionshub.epam.com/blog/post/idea-generation-in-business (accessed January 16, 2024).

11. Chau JPC, Chien WT, Liu X, Hu Y, Jin Y. Needs assessment and expectations regarding evidence-based practice knowledge acquisition and training activities: A cross-sectional study of healthcare personnel in China. *Int J Nurs Sci* 2022;9:100–6. https://doi.org/10.1016/J.IJNSS.2021.11.001.
12. Gorgon E, Maka K, Kam A, Nisbet G, Sullivan J, Regan G, et al. Needs assessment for health service design for people with back pain in a hospital setting: a qualitative study. *Heal Expect* 2022;25:721–31. https://doi.org/10.1111/HEX.13419.

Chapter 4

Empathy and Personas: Stepping into Customers' Shoes and Understanding Medical Issues

4.1 Understanding the Needs of a Customer and Patient

It is well known that a startup or an entrepreneurial journey which fails to understand its customer eventually fails. Even if the quality of their product or service is premium or superior to other ventures, a startup can fail if they lack in understanding how a customer thinks and perceives. These tend to the popular saying "Stepping into customers' shoes," meaning that an entrepreneur should first think from the perspective of a customer and then design their products accordingly [1]. In the healthcare domain, in many cases, the term customer is usually replaced by "patient." Hence, it becomes necessary to truly embrace the behavioural marketing while understanding the patients empathetically [2].

Lack of empathy and understanding of patient personas could lead to the failure of startups and new ventures. Hence, the primary objective of a healthcare entrepreneur is to empathetically understand the needs of the patient for which they wish to build the products or services. In today's world, even the most popular medical product is not featured as a high

DOI: 10.4324/9781003475309-6

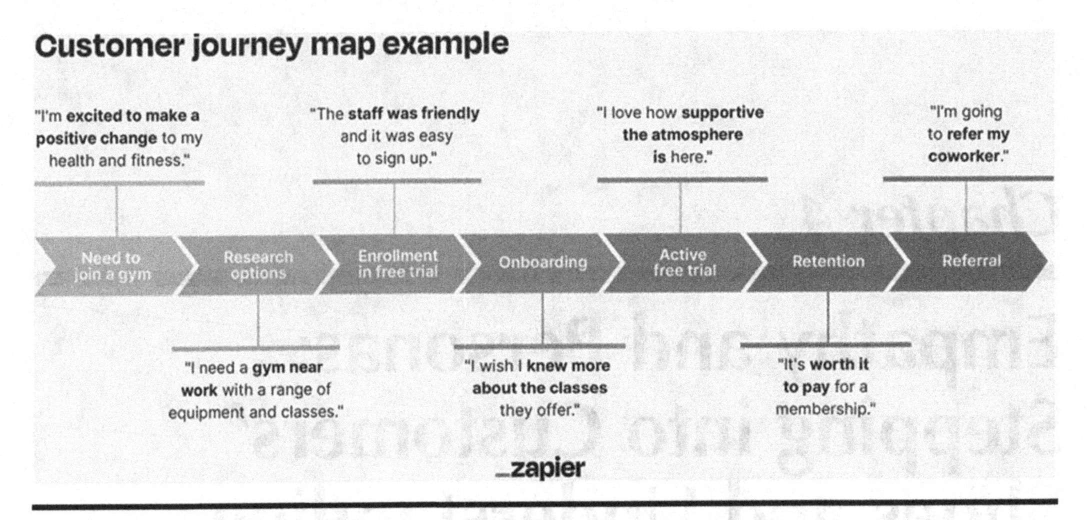

Figure 4.1 An illustration representing a customer's journey [6].

margin product. Rather, healthcare entrepreneurs optimise the products, services, and their marketing with respect to the patient's journey; hence, mapping a patient's pain journey is of utmost importance to the design of any new medical device or service [3].

By adopting a patient-centric approach, the traditional entrepreneurial situation may be altered by placing greater emphasis on the patients and engaging them more actively in their own care procedures. A patient-centric approach, for instance, enables individuals to collaborate with clinicians in order to comprehend their own health risks and treatment alternatives, as well as to actively participate in the decision-making process. The potential advantages of this methodology can be substantial [1, 4, 5]. Figure 4.1 shows an illustration of a customer's journey.

Although the advantages of adopting a patient-centric approach are evident, several businesses are now unprepared to facilitate consumer participation in healthcare decision-making. Furthermore, several professionals fail to see the need for devoting time to educating patients on how to be more proactive with their health. This necessitates that healthcare managers evaluate their own organisational capabilities and determine the most effective ways of advancing patient-centred initiatives [7]. Effectively doing this necessitates that the company and its practitioners possess a more profound comprehension of how engaging the customer may contribute to overarching organisational and public health objectives. It is crucial to acknowledge that not all patients are the same when formulating a patient-centric strategy for an healthcare entrepreneur. Consequently, the manner the entrepreneur takes towards this difficulty may vary based on which individuals are targeted. An entrepreneur may establish productive working collaborations by more precisely customising the marketing and outreach initiatives in accordance with their target audience.

For comprehending the variety of patients a healthcare industry caters to, a technique to employ a matrix including four distinct sections is usually used. In this, within one section are those who are physically fit and conscientious about their own well-being. They may adhere to suggested health screening criteria, engage in regular physical activity, consume a well-balanced diet, and utilise an activity monitoring gadget such as a Fitbit. Another section has individuals who are ill yet do not take responsibility for their own health. Individuals situated in this particular section may be afflicted with several chronic ailments, rendering their medical treatment presumably rather costly. The third section consists of healthy but uninvolved individuals, such as Millennials, who would sometime take risks, disregard the necessity of investing in health insurance and treatment because they are unaware of the circumstances. The last section comprises individuals who are ill but actively participate in health maintenance practices, such as a heart attack survivor who now maintains a healthy diet and daily walking regimen. After identifying the target demographic or populations, an entrepreneur must also consider how they will engage them and attract them to their network. Figure 4.2 shows the summary of patient segments an entrepreneur may consider to select a best fit for their product or service. In addition to this, customer experience related to a service or product should be understood effectively.

According to Dr John Quelch [6], the following are the five pointers which result in good healthcare product or service experience for patients or customers:

1. Product or service experience leading to cure.
2. Having empathy to understand them and make them feel that the service provider cares for them.
3. Efficiency in resolving customer complaints.
4. Better prices to ensure fairness and competition with the market.
5. Empowering patients to provide choices in selecting their service plans.

In summary, inquiry should be conducted with a provider of market or consumer research that can furnish the necessary facts to make informed selections regarding your patient's or customer's desires and requirements. An entrepreneur cannot genuinely devise an effective plan for moving forward

Healthy Segment
Eats healthy, involved in taking care of themselves, use activity trackers, eat balanced diet, goes for regular health screenings
Sick and Unhealthy Segment
Eats unhealthy, not involved in taking care of themselves, unwilling to become engaged, healthcare is expensive
Uninvolved Segment
They are healthy but uninvolved, such as Millennials, who feel they are invincible and so don't see the need to invest money on health insurance and treatment.
Sick but involved segment
Individuals who are ill yet actively participate in maintaining their health

Figure 4.2 The patient segments an entrepreneur may consider to select the best fit for their product or service.

and fostering a deeper relationship that benefits both customers and the company until they have all the relevant information.

4.2 Techniques for Empathy and Persona Mapping

4.2.1 Empathy Mapping

The capacity to relate to the feelings of others is empathy. Acquiring this talent is crucial in the job. Demonstrating empathy towards others is likely to elicit reciprocal responses, so facilitating cooperation, collaboration, and teamwork [8]. Prospective clients, peripheral stakeholders, and distant employees are those with whom one seldom interacts. Occasionally, it might be challenging to empathise with their situation. One may gain a deeper understanding of how these individuals truly feel and think about your product, service, or circumstance by employing "Empathy Mapping" [9]. The empathy map was first developed by Dave Gray at Xplane [10] with the intention of reducing miscommunication and misunderstanding about target

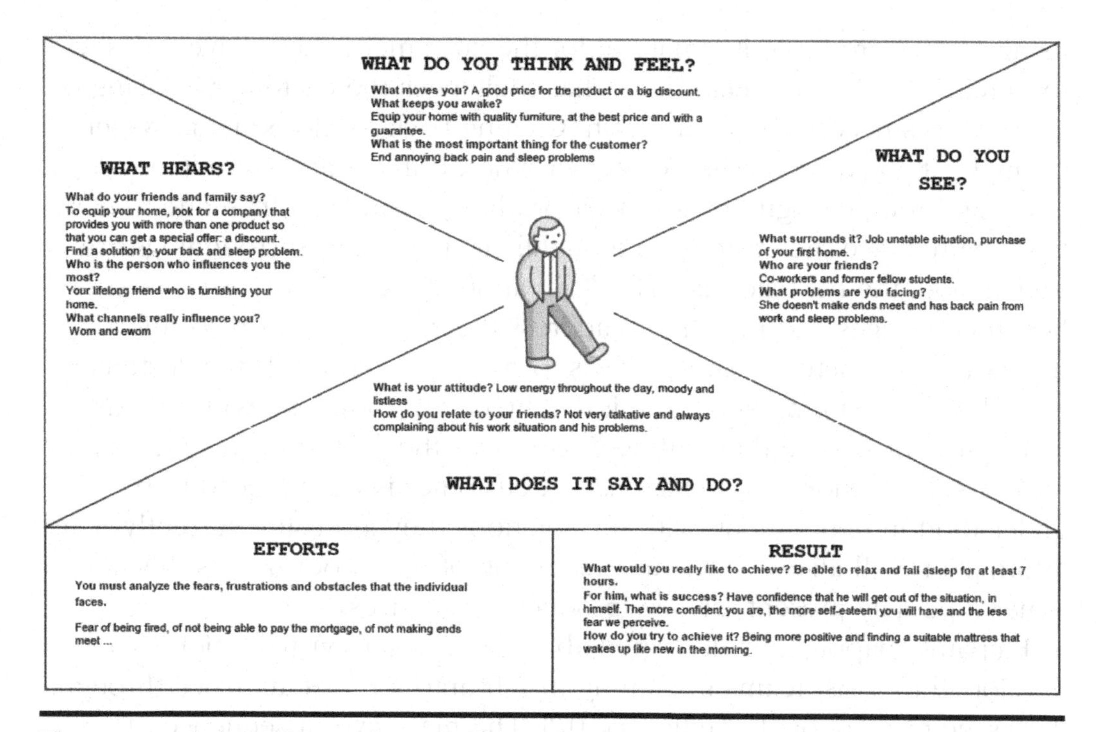

Figure 4.3 A typical empathy map illustration [11].

groups, including consumers and users. A typical empathy map illustrated by Illescas-Manzano et al. [11] is shown in Figure 4.3.

In an empathy map, as shown in Figure 4.3, information is first obtained on what a customer hears or gets influenced by. It could be social media, family or friends which make them aware or conscious about any healthcare challenge they may be facing. The next step is to analyse what they think or feel. This includes product features which move them, such as a good price for the product or a big discount. Also, what keeps the customer awake or motivated needs to be investigated. An example is their desire to purchase quality products, at the best price and with a guarantee. Additionally, the most important thing for the customer needs to be determined. Specifically, it has to be ascertained which health issue or challenge bothers them the most. Beyond what a customer thinks or feels, it is important to understand what they see. Similar to the examples in Figure 4.2, is there a particular health issue they had been facing or ignoring for a long time, and desiring to purchase or own a particular product which they cannot afford. An expensive body massager could be an example which a middle-aged mother desires due to prolonged body pain but just can't afford due to limited salary. It is also imperative to understand who are the customer's friends and colleagues. If they are using certain products, it is natural for the customer also to have a desire to purchase such products. Also, what challenges the customer is facing on a daily basis need to be focused on. Coming back to the body massager example, it could be that the customer who desires it and cannot afford is going through a significant lack of productivity in her office due to her body pain. Once the customer's thoughts, feelings, and surroundings are understood, the next step is to find out about their actions. Given their health challenges and inability to address them, the level of discomfort the customer is actually facing needs to be accounted for. If the customer is further complaining about the issue directly or indirectly with friends, colleagues, or people they talk to, it confirms their distress. In summary, the fears, frustrations, and challenges a customer is facing, need to be understood in depth along with the solutions they are using currently. This information will allow better understanding of the opportunities this customer type may present for new products or services.

Empathy mapping, as described above, is a potent visualisation tool developed to assist teams in gaining insight into a target audience through the usage of emotional intelligence [12]. The map gives a sequence of prompts for determining the ideas, emotions, motivations, wants, and requirements of a patient or a customer [13]. This compels the investigative team to prioritise the needs of the target audience over its own. An example of how a healthcare entrepreneur may utilise an empathy map is to contemplate how patients would react to a novel gadget or a source of discomfort. One may be utilised by a team manager to evaluate the team's response to a novel process. Because empathy maps are extremely visual, they are simple to distribute and use to identify knowledge gaps and research discrepancies. Equipped with this knowledge, groups may produce items or supply solutions that are more efficacious and in accordance with consumer demands.

As empathy mapping can be a group activity, ensuring that all participants are present in the same room is important. Providing a sufficient quantity of Post-it ® notes and marker pens can enable everyone to participate. When working with a bigger group, utilising a whiteboard that has been imprinted or sketched with a huge, empty empathy map may be beneficial. Printing a page for a smaller group and have each member annotate it, leaving sufficient space around the diagram's important components would be useful. Figure 4.4 shows a simplified empathy mapping template. The following key steps may be followed to generate the empathy map as per template suggested by the Nielsen Norman Group [14].

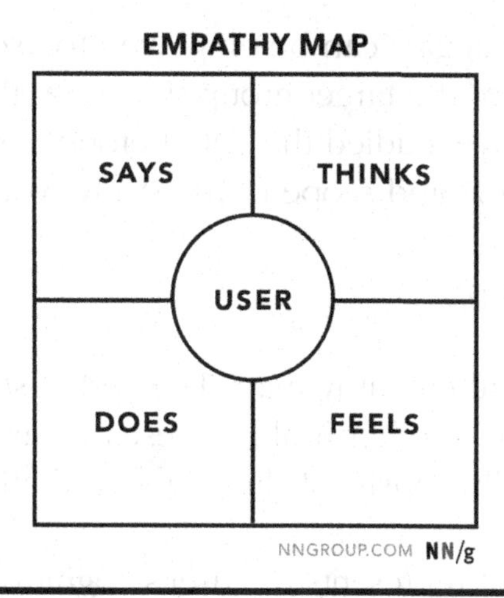

Figure 4.4 Empathy map template [14].

1. Determine the scope and subject of your empathy map.

An entrepreneur, for instance, can map a client requesting the development of a new healthcare service application or a test group of patients. One potential approach is to utilise personas, which are fictitious archetypes of customers or stakeholders, to symbolise these groupings. What precisely must an entrepreneur's empathy mapping session accomplish or perform? The responses will determine how the subsequent steps of the mapping procedure will go. The utilisation of customer experience mapping may prove advantageous in identifying and focusing on particular "touchpoints." Recording here any specific inquiries or requests that the stakeholder may have had regarding the topic of examination would help [7].

2. Collection of relevant data is important.

Empathy mapping is most effective when it is constructed upon the genuine ideas, emotions, and perspectives of its participants. It is advisable to gather a variety of data on the experiences of stakeholders as an integral component of the development process. For instance, they may be derived via subject interviews, surveys, observation, or the analysis of recorded sales calls. Demonstrating inquisitiveness and employing appropriate questioning techniques to elicit the necessary information is the key. When conducting in-person interviews, it is important to carefully listen and observe non-verbal clues in order to gain a more complete understanding of the interviewee's thoughts and emotions. When conducting subject interviews is not feasible, market segmentation strategies can be employed to ascertain the characteristics and requirements of the target group. Ensuring that all participants in the mapping exercise have studied the data beforehand and have a thorough understanding of the topic and scope of the study would be helpful.

3. Fill the empathy map.

Figure 4.4 shows a typical empathy map. The verbal statements made by the user during an interview or other usability research are shown in the "Says" quadrant. It should ideally consist of direct and verbatim quotations from scholarly sources.

The "Thinks" quadrant represents the user's cognitive processes during the course of the interaction. Based on the qualitative study that was collected, one should inquire about the user's preoccupations. What is essential to the user? The identical material might potentially be included in both "Says" and "Thinks." However, particular consideration may be required for the opinions of people who may be reluctant to express them. It is important to make an effort to comprehend their reluctance to divulge; are they uncertain, self-aware, courteous, or apprehensive about telling people something? The user's activities are contained inside the "Does" quadrant. According to the research, what is the user's physical activity? How does the user proceed with the task? In the "Feels" quadrant, the user's emotional state is frequently denoted with a brief statement supplemented with an adjective to provide context. It is important to consider: What concerns the user? What causes the user to get enthusiastic? What is the user's opinion on the experience? [14].

Certain quadrants may appear unclear or redundant; for instance, differentiating between Thinks and Feels could prove challenging. Excessive emphasis on precision should be avoided; if an object can be accommodated in numerous quadrants, one of them needs to be selected. The four quadrants exist just to reinforce the understanding of users and to guarantee that no vital dimension is overlooked.

4. Conclude.

If the entrepreneur requires further elaboration or possesses distinct requirements, they may modify the map by including further quadrants or by augmenting the specificity of existing quadrants. Purpose-specific refinement and digitisation of the output should be applied to the empathy map. Points to be mentioned include the username, the version number, any lingering questions, and the date. Revisiting the empathy map iteratively may be useful in the process of gathering further research or to inform user experience decision-making. Creating a value proposition for the product which the entrepreneur is in the process of building or for the issue they are attempting to resolve would be the next step. Empathy mapping may also be employed as an initial phase in design thinking to facilitate the delivery of patient-centred solutions.

4.3 User Personas

User personas are customer avatars, "pen portraits," or "archetypes" of the individuals an entrepreneur wishes to convince to purchase their product or service. They serve as portraits of the present and prospective clients, showcasing their respective behaviours [15]. Personas may illustrate how, where, and when a consumer could interact with a product, as well as the reasons why that customer would be interested in it. Personas are not founded on assumptions or expectations but rather on study and evidence, and they contain more than simply facts.

Personas serve as a valuable tool for addressing knowledge gaps regarding prospective patients. They comprise a blend of qualitative and quantitative data that can be leveraged to optimise many aspects, including the website and customer acquisition strategies. In the absence of personalities, every aspect of the marketing strategy is founded upon mere conjecture and presumption. As the pursuit of enhancing customer experiences gains momentum, the most effective course of action is to exhibit to prospective customers as individuals with their aspirations, anxieties, incentives for seeking care, and the process they employ to determine which provider to engage with. Figure 4.5 shows the framework used by Rohmiyati et al. [16].

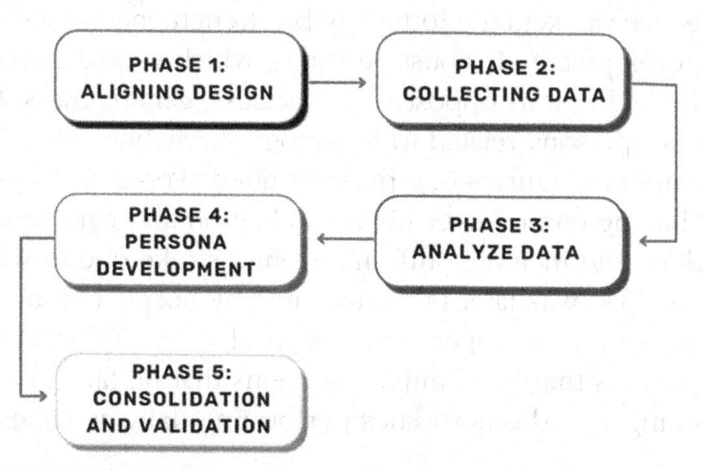

Figure 4.5 Framework implemented to generate the personas [16].

Per-Olov Larsson

67 years old

Per-Olov had a stroke 4 years ago. He came to the hospital quickly and got treated with Thrombolysis. After getting discharged from hospital, Per-Olov received rehabilitation at a neurological rehabilitation clinic. He was then discharged to home and received support from a neurology team for continued rehabilitation. Per-Olov lives with his wife Lean and has 3 adult children. Previously, he was working as a teacher but has not returned to work after his stroke.

"I want to be able to follow what is happening in healthcare and get an overview of the opportunities and choices I have"

Health problems
Diabetes type 2

Hypertension (treatment by medication since 15 years ago)

Background
Per-Olov suffered a stroke 4 years ago. He woke up early one morning and fell down on his way to the bathroom. His wife, Lena, realized that something was wrong and called an ambulance.

Motivation
Per-Olov is engaged in the local stroke association and has a good insight into the problems that many stroke patients experience after stroke.

Problem description
Per-Olov still suffers from disabilities after stroke. He has concentration difficulties and suffers from fatigue. His fatigue stops him from doing many things in his daily life.

Needs
Per-Olov is primarily interested in having control over his care and rehabilitation. He wants to know which alternatives he has in choosing a care provider and which support they do provide. It is of great importance for him to find the right caregiver that provides the support that he needs.

Figure 4.6 A patient's persona related to neurological rehabilitation [17].

Personas serve as means to articulate the thoughts, motivations, and purchasing patterns of certain customers. They provide direct engagement and influence over individuals' decision-making processes, potentially leading to increased sales and even a willingness to pay a premium price in the long run. Effective marketing efforts may be strengthened and concentrated through the development of robust personas, which enable communication with "actual individuals" as opposed to a faceless, generic mass. Figure 4.6 shows a patient's persona related to neurological rehabilitation.

In the example in Figure 4.6, a persona titled "Per-Olov Larsson" is interested in having control over his rehabilitation and care schedules. His personal background includes suffering from a stroke due to which he was left unconscious. He was later taken to a nearby hospital by ambulance. His motivation is to get engaged with the local stroke association and guide other patients that face similar problems that he faced post stroke. He still faces fatigue and other issues performing daily activities due to disabilities post stroke. Moreover, he suffers from type 2 diabetes and chronic hypertension. This persona describes a larger chunk of population in their late 50s or 60s who have left their job due to these disabilities. These types of information can help an healthcare entrepreneur target a particular population in providing appropriate services or healthcare products in enhancing the overall lifestyle of such patients. Hence, rather than performing traditional marketing approaches, an entrepreneur may focus on a converged target.

By shifting from conventional marketing approaches to campaigns that are consistent, targeted, and individualised, an entrepreneur may establish a connection with clients, employ their language, and directly attend to their requirements. Furthermore, by being exact in their marketing efforts, it will avoid squandering time on unsuccessful sales attempts [18]. The following steps can be followed to develop personas for customers with healthcare needs or for patients.

1. Commencing with the One Person: The individual for whom the most critical service line is intended. We need to proceed with the understanding that we have a certain service line or area of expertise that we are primarily focused on providing content for. This service line will serve as the foundation for the content marketing campaign that is to be constructed. Following that, rinse and repeat is the way to go, followed by developing connecting material for other service lines.
2. Conducting interviews with current clients of the preferred service line by telephone or in person. The focus during patient interviews on their health journey should be centred around their issue and the manner in which a resolution may be provided.
3. Background: It is important to understand the fundamentals of the function that an individual and their family play in the healthcare system. Recording the essential details pertaining to their residence and the way of life of their family is important. Furthermore, questioning may be conducted about the health of the individual and their family.
4. Demographics: These should contain information about gender, age, family income, whether the household is located in an urban or rural area, which websites do people consult for health-related inquiries and which digital channels do they commonly utilise for entertainment or personal interaction.

5. Psychographics: This section contains information about the customer ethnicity, culture, interests, food choices, eating habits, and travels.
6. Goals, Challenges, Objections, and Content: This should mention the persona's primary, secondary, and tertiary health objectives and challenges. Major objections could include availability of specialists, insurance claims, and appointment delays. Content could include the exact information the persona is looking for and how that solution should be presented. Figure 4.7 shows a sample patient's persona.

In the example in Figure 4.7, a persona called "Monika" is a 65-year-old woman, who is a musicologist and is in transition to retirement. Her personal background includes working with a church and as a lecturer at a university. She lives in a large rented apartment in a metropolitan city. She lives

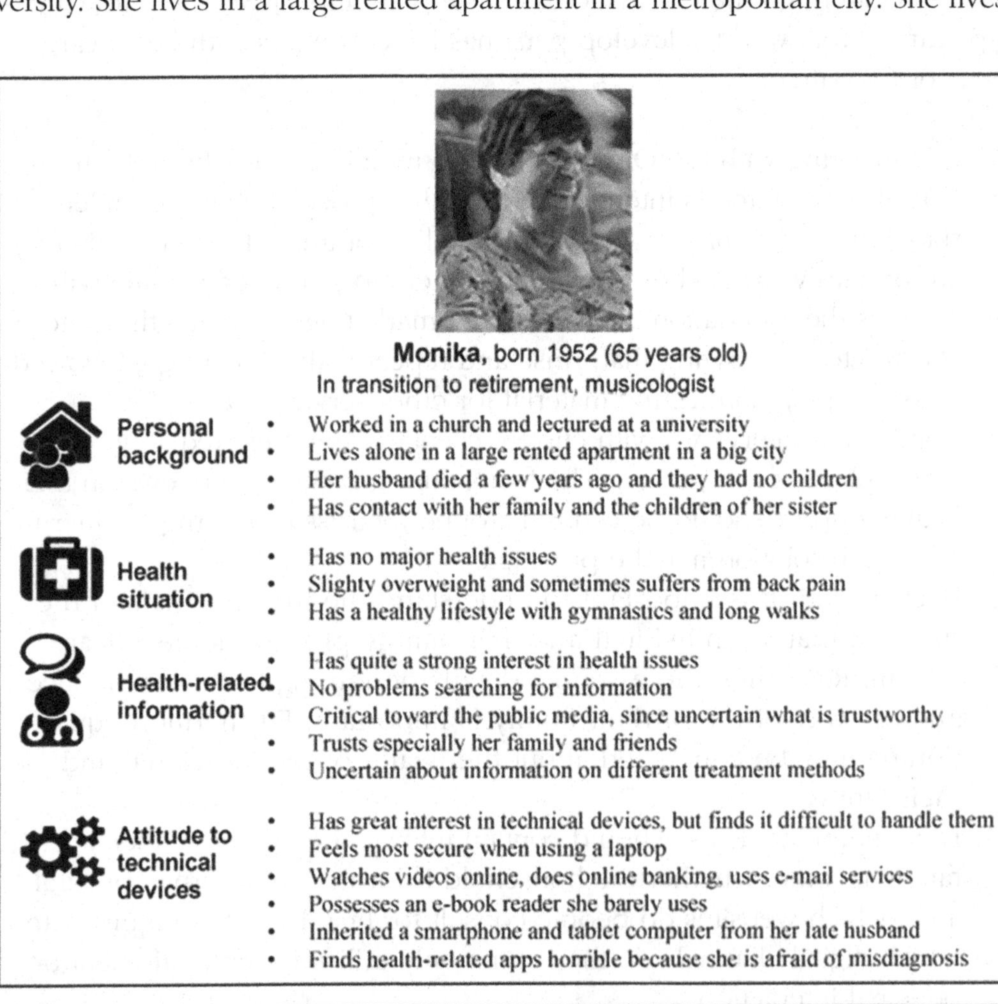

Monika, born 1952 (65 years old)
In transition to retirement, musicologist
Personal background
Worked in a church and lectured at a university
Lives alone in a large rented apartment in a big city
Her husband died a few years ago and they had no children
Has contact with her family and the children of her sister
Health situation
Has no major health issues
Slighty overweight and sometimes suffers from back pain
Has a healthy lifestyle with gymnastics and long walks
Health-related information
Has quite a strong interest in health issues
No problems searching for information
Critical toward the public media, since uncertain what is trustworthy
Trusts especially her family and friends
Uncertain about information on different treatment methods
Attitude to technical devices
Has great interest in technical devices, but finds it difficult to handle them
Feels most secure when using a laptop
Watches videos online, does online banking, uses e-mail services
Possesses an e-book reader she barely uses
Inherited a smartphone and tablet computer from her late husband
Finds health-related apps horrible because she is afraid of misdiagnosis

Figure 4.7 A sample patient's persona by Schäfer et al. [19].

alone as her husband died a few years back and they had no children but is in contact with her family. Her health situation includes being overweight and sometimes suffering from back pain. She has a healthy lifestyle as she goes to the gym and takes long walks. She has interest in reading about health issues and faces no problem in searching for the required information. Although she has great interest in new technology, she faces problem in handling it. She is also sceptical about healthcare applications as she is afraid of wrong treatment or misdiagnosis.

This persona describes a wide range of elderly female customers who live alone and rely on technology for healthcare information. The information can help in identifying entrepreneurial opportunities such as with healthcare products or services which are aimed at providing reliable advice through the mouths of verified doctors or certified medical professionals. Hence, development of personas is important to understand the requirements of a patient.

Once the personas are developed, as observed in the example above, they will provide valuable insights for designing and developing products, as well as for informing paid marketing endeavours, website content, and social media material from a marketing standpoint. Potentially beneficial discoveries pertaining to the new patient intake process or even strategies for answering phone calls may be discovered internally. Contemporary patients desire more intimate rapport with their healthcare professionals. By providing insights beyond the patient's medical records, personas enable us to develop an unparalleled patient experience from the moment they visit a website until their final office visit. In light of the newly acquired customer segmentation, as an entrepreneur, it is important to consider if they are able to effectively communicate with each persona. If they happen to have been too skewed in one direction, by utilising personas as a reference point, they may strategically realign their messaging and marketing efforts in order to effectively engage with each demographic.

In another example persona (Figure 4.8), Melissa is a new expectant mother who, like other women, wants to have a healthy baby. It was hard for her to get pregnant and she is worried that she will lose the baby. The reason for her worry is due to her traumatic past in which she had experienced a miscarriage. She wants an empathetic doctor who understands her situation and considers her health condition seriously. She wants advice on staying healthy during the course of her pregnancy. This persona reflects the possible need for certified and empathetic medical professionals which

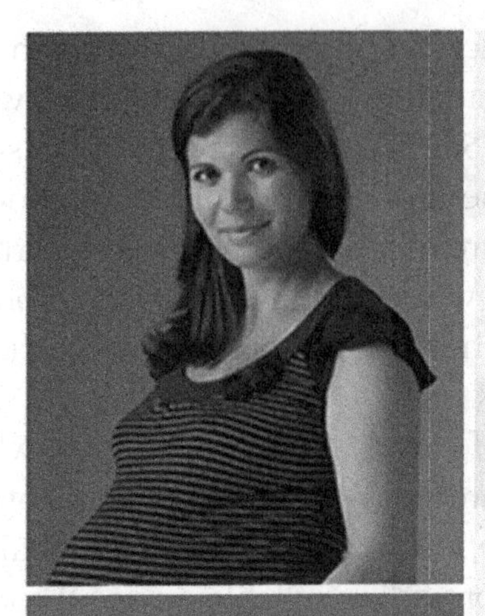

Melissa
NEW EXPECTING MOTHER
Age: 34
Occupation: Executive Assistant
Education: Some College
Relationship: Married
Children: None
"It's been hard for us to get pregnant and I'm so worried I'll lose this baby."
Currently I feel...
Happy Concerned Stressed Busy
My healthcare goals
I want to have a healthy baby. It's been our dream for a year. I want to work with a doctor who understands how worried I am about this pregnancy and takes me seriously. I want to know about the risks of miscarriage during my pregnancy. I want to feel like I can call my doctor with concerns. My doctor should help me know what I should be doing to stay healthy during my pregnancy.
About me
I'm so happy to be pregnant, but very worried because I had a previous miscarriage. I've heard that there are more risks of complications for women who have their babies at my age. I also know I'll need to work very hard to keep this baby from coming prematurely. I expect I'll be put on bed rest later in my pregnancy.

Figure 4.8 Another example related to a pregnant patient's persona [20].

could guide her to stay healthy during her pregnancy, and this information could be used by entrepreneurs to develop or market products to their target audience.

4.4 Empathy Mapping and Persona Maps: Differences

While both empathy mapping and persona maps may appear close, if not identical, upon initial inspection, there are significant distinctions between the two. It is worth examining the nature of the information employed in each. User interviews serve as the information source for empathy maps [21]. To produce a high-quality empathy map, direct encounters with actual users are required. A persona, conversely, obtains its information from interviews, surveys, and databases; hence, unlike empathy maps, personas need the gathering of quantitative data and a substantial quantity of different material. From the viewpoints of each individual, empathy maps are more specific in that they depict user characteristics in certain situations and environments. Empathy maps depict a certain moment in the user's life and are motivated by the user's thoughts, actions, emotions, and feelings at that time. In contrast, user personas prioritise the holistic understanding of the user. User personas gather information that aids in the definition of the user; "Who is the user? What is their type?" serves as the essential questions for personas. User personas employ a comprehensive approach, adopting a fisheye view, and endeavour to depict the user completely by using precise demographic and psychographic data [22].

The fundamental distinction between an empathy map and a persona is that the former is constructed based on information from actual individuals and the latter is a broad understanding of fictitious users. User personas identify your target audience, whereas empathy maps depict the attitudes and behaviours of those individuals [23]. Momentary human emotions have the potential to significantly impact an empathy map. For instance, a user's reaction may vary depending on their emotional state—for instance, sadness or happiness. User personas remain unaffected by this factor. Generally, empathy maps are constructed with the content of a persona's thoughts, actions, and emotions in mind. Consequently, an empathy map may be seen as a subset of personas that is concerned with a certain domain. In conclusion, personas and empathy maps cannot substitute for one another in terms of understanding consumers' requirements and demonstrating empathy, but are equally important for understanding the customers.

References

1. Fisk P. *Walking in the Customers' Shoes. Mark.* Genius, John Wiley & Sons, Ltd; n.d. https://doi.org/10.1002/9780857086518.CH9.
2. Purmonen A, Jaakkola E, Terho H. B2B customer journeys: conceptualization and an integrative framework. *Ind Mark Manag* 2023;113:74–87. https://doi.org/10.1016/J.INDMARMAN.2023.05.020.
3. Lin LZ. Modeling and analysis of customer journey enablers: a case study of religious pilgrimage. *J Hosp Tour Manag* 2023;57:200–12. https://doi.org/10.1016/J.JHTM.2023.10.004.
4. Burtonshaw-Gunn SA. Customer relationship management. *Essent Manag Toolbox* 2012:45–54. https://doi.org/10.1002/9781119208723.CH3.
5. Walters D. *Customer Journey Mapping: Putting Yourself in the Customer's Shoes. Behav.* John Wiley & Sons, Ltd; 2015, pp. 69–82. https://doi.org/10.1002/9781119170440.CH07.

6. Customer journey mapping 101 (+ free templates) | Zapier n.d. https://zapier.com/blog/customer-journey-mapping/ (accessed November 24, 2023).
7. Empathy mapping – understanding your customers' thoughts and feelings n.d. https://www.mindtools.com/abtn3bi/empathy-mapping (accessed November 24, 2023).
8. Marsden N, Wittwer A. Empathy and exclusion in the design process. *Front Hum Dyn* 2022;4:1050580. https://doi.org/10.3389/FHUMD.2022.1050580.
9. Empathy map the definitive guide: creating 10-minute user persona n.d. https://www.uxpin.com/studio/blog/the-practical-guide-to-empathy-map-creating-a-10-minute-persona/ (accessed November 24, 2023).
10. Updated empathy map canvas. We designed the empathy map at XPLANE... | by Dave gray | Medium n.d. https://medium.com/@davegray/updated-empathy-map-canvas-46df22df3c8a (accessed November 24, 2023).
11. Illescas-Manzano MD, López NV, González NA, Rodríguez CC. Implementation of Chatbot in online commerce, and open innovation. *J Open Innov Technol Mark Complex* 2021;7:125. https://doi.org/10.3390/JOITMC7020125.
12. Cairns P, Pinker I, Ward A, Watson E, Laidlaw A. Empathy maps in communication skills training. *Clin Teach* 2021;18:142–6. https://doi.org/10.1111/TCT.13270.
13. Pileggi SF. Knowledge interoperability and re-use in empathy mapping: an ontological approach. *Expert Syst Appl* 2021;180:115065. https://doi.org/10.1016/J.ESWA.2021.115065.
14. Empathy mapping: the first step in design thinking n.d. https://www.nngroup.com/articles/empathy-mapping/ (accessed November 24, 2023).
15. The power of patient personas for physician practice marketing – Carenetic digital n.d. https://careneticdigital.com/the-power-of-patient-personas-for-physician-practice-marketing/ (accessed November 24, 2023).
16. Rohmiyati Y, Tengku Wook TSM, Sahari N, Hanawi SA, Qamar F. Designing personas for e-resources users in the university libraries. *Comput* 2023;12:48. https://doi.org/10.3390/COMPUTERS12030048.
17. Davoody N, Koch S, Krakau I, Hägglund M. Post-discharge stroke patients' information needs as input to proposing patient-centred eHealth services. *BMC Med Inform Decis Mak* 2016;16:1–13. https://doi.org/10.1186/S12911-016-0307-2/FIGURES/4.
18. Developing personas – ribbleg your buyers to "get closer" to them n.d. https://www.mindtools.com/a9rsycl/developing-personas (accessed November 24, 2023).
19. Schäfer K, Rasche P, Bröhl C, Theis S, Barton L, Brandl C, et al. Survey-based personas for a target-group-specific consideration of elderly end users of information and communication systems in the German health-care sector. *Int J Med Inform* 2019;132:103924. https://doi.org/10.1016/J.IJMEDINF.2019.07.003.
20. Persona design for healthcare company by Lauren Shirrell on Dribbble n.d. https://dribbble.com/shots/2573226-Persona-Design-for-Healthcare-Company (accessed November 24, 2023).

21. Empathy Map vs Persona: what's the difference and why you need both n.d. https://userpilot.com/blog/empathy-map-vs-persona/ (accessed November 24, 2023).
22. Empathy maps vs user personas, what is the difference? | by Hussein Najah | Bootcamp n.d. https://bootcamp.uxdesign.cc/empathy-maps-vs-user-personas-what-is-the-difference-901434d0262c (accessed November 24, 2023).
23. Personas and empathy mapping for understanding customers and users – the persona blog n.d. https://persona.qcri.org/blog/personas-and-empathy-mapping-for-understanding-customers-and-users/ (accessed November 24, 2023).

Chapter 5

Effective Patient Survey and Understanding Market Competition

5.1 Effective Patient Survey

Patient satisfaction serves as a metric gauging the contentment and overall happiness of individuals with their healthcare experiences. This encompasses various elements, including the quality of care received, communication with healthcare providers or physicians, accessibility of services, and the overall experience within hospitals or clinics [1]. The patient's journey, spanning appointment booking, waiting times, interactions with healthcare professionals, engagement with hospital staff, and the billing process, significantly influences patient satisfaction.

A patient satisfaction survey functions as a structured set of inquiries designed to gather feedback from patients, systematically evaluating their contentment with the quality and care provided by healthcare service providers [2]. This questionnaire plays a pivotal role in assessing fundamental metrics related to patient care, enabling medical institutions to comprehend the extent of care delivered and identify potential shortcomings in service delivery.

The primary points to keep in mind while performing or modelling a survey are as follows:

- Keeping it short: The primary objective is to formulate questions with clarity and conciseness, eliminating superfluous phrasing without compromising the intended meaning. While the reduction of character

 DOI: 10.4324/9781003475309-7

count is paramount, it is equally essential to uphold a reasonable survey length to minimise the likelihood of participant abandonment. Reflecting on the improbability of participants engaging enthusiastically in an extensive 30-minute questionnaire would be valuable.

- Only ask questions that fulfil the end goal of an entrepreneurial venture: In essence, adopt a stringent approach in excising redundant queries from the surveys. Each question should serve a well-defined purpose, substantiating its inclusion convincingly; otherwise, it warrants reconsideration for removal [3].

  For example, contingent upon the survey's objectives, the method by which a customer initially interacts with the site may lack relevance. In such instances, abstain from seeking information about their source of discovery. Similarly, if the acquisition of a customer's name proves non-essential, abstain from positing the question.

  The inclusion of ostensibly innocuous questions has the potential to needlessly prolong the survey, potentially prompting respondents to navigate towards the "back" button. Hence, a judicious selection of questions ensures the efficiency and brevity of the survey instrument.
- Construct smart, open-ended questions: While the allure of relying on multiple-choice queries and scales is strong, it is important to recognise that some of the most insightful feedback emerges from open-ended questions. These questions afford customers the opportunity to express their authentic thoughts freely.

  However, the inclusion of a substantial text box linked to the initial question can be overwhelming, potentially deterring survey participants. It is prudent to commence the survey with concise questions to establish a sense of progress. Subsequently, for participants who have reached the concluding questions, providing them with the chance to expound on their thoughts becomes more conducive.

  One effective approach involves prompting respondents to commit to a question through a straightforward introduction, followed by an open-ended query such as, "Why do you feel this way?" This method encourages thoughtful responses and enhances the overall depth of feedback obtained.
- Ask one question at a time: Encountering an extensive series of inquiries, such as "How did you find our site?" "Do you understand what our product does?" "Why or why not?" can create a sense of being interrogated, impeding respondents from expressing their thoughts comprehensively. To elicit responses of substance, it is imperative to allocate ample time for individuals to contemplate each question independently.

The simultaneous presentation of multiple questions often results in perfunctory responses, as respondents may be inclined to expedite the process or, regrettably, abandon the survey prematurely. A more efficacious approach involves optimising the survey structure by concentrating on one key point at a time. This not only facilitates more thoughtful and comprehensive responses but also enhances the overall quality of feedback obtained from participants. Such a methodical approach contributes to a more meaningful and insightful survey experience.

- Make rating scales consistent: Common survey scales can pose challenges and create confusion when the context undergoes changes. Consider this example: In the initial survey questions, respondents are instructed to choose between 1–5, where 1 signifies "Strongly Disagree" and 5 denotes "Strongly Agree."

  However, as the survey progresses, participants may be tasked with evaluating the importance of certain items, and the scale undergoes a shift. Now, 1 is designated as "Most Important," creating potential confusion, especially for those who had consistently used 5 as the agreeable response in preceding questions. This transition introduces a considerable risk of respondents inadvertently providing inaccurate answers due to the altered scale, highlighting the need for clarity and consistency in survey design to ensure accurate and meaningful responses.
- Avoid leading and loaded questions: Survey questions characterised by biased phrasing, which subtly guides respondents toward a predetermined answer, pose a significant challenge as they compromise the integrity and accuracy of the feedback obtained. A notable example from SurveyMonkey exemplifies the importance of avoiding leading questions: "We have recently upgraded SurveyMonkey's features to become a first-class tool. What are the thoughts on the new site?"

  This instance underscores the difficulty of maintaining objectivity when expressing pride in a product. A more impartial alternative, such as "What do you think of the recent SurveyMonkey upgrades?" ensures a fairer approach to soliciting feedback.

  In the pursuit of genuine insights, it is imperative to refrain from employing leading questions or other tactics designed to manipulate responses. Such practices not only risk causing frustration among participants but also introduce bias, thereby compromising the reliability and validity of the collected data.

- Get specific and avoid assumptions: Specific subjects introduce complexities, particularly when aiming to capture responses from a diverse audience. A substantial hurdle arises from the use of industry acronyms, buzzwords, jargon, or references in questions, prompting the recommendation to abstain from such language.

  Making assumptions about respondents' inclination to provide specific examples or elucidate their reasoning poses challenges. A more effective approach involves explicitly encouraging specificity and conveying an openness to comprehensive feedback. For instance: "How do respondents perceive? They are encouraged to furnish specific details, as we value detailed and thorough feedback."

  By adopting a more inclusive methodology and refraining from presumptions about the extent of respondents' knowledge, surveys can elicit more accurate and meaningful responses from a broad spectrum of participants.
- Offer survey respondents a bonus: In certain scenarios, it becomes prudent to encourage customer participation in surveys, as evidenced by various data indicating that incentives can effectively augment survey response rates [4]. These incentives may manifest in the form of discounts, giveaways, or account credits.

  A pivotal aspect lies in finding a harmonious balance where customers find sufficient motivation to engage in the survey without unduly straining the financial resources of the brand. Opting for incentives that align with the brand's fiscal capacity, such as credits or free trials, is frequently recommended over unrelated gifts or substantial discounts. While concerns may arise regarding the potential impact of offering free incentives on the quality of responses, empirical studies suggest that such apprehensions are likely unwarranted.

A comprehensive patient satisfaction survey questionnaire is imperative for the systematic collection of feedback [5]. This exemplar survey can be tailored to incorporate specific details mandated by relevant authorities. The following five example questions could help for inclusion in a patient satisfaction survey [6]:

Q1: Derived from the overall experience with our medical care facility, kindly indicate the probability of recommending our services to a friend or colleague

Healthcare has undergone substantial evolution, transforming into an expansive industry. Trust is paramount, particularly in the context of seeking medical care [7]. Among various industries, healthcare relies significantly on recommendations from past or current patients, influenced by the quality of care and resulting satisfaction.

Q2: Were there any challenges encountered in scheduling an appointment?

The prompt allocation of appointments holds immense significance for patients, considering their health conditions. Timeliness in this regard is crucial, as individuals who are unwell or suffering from ailments should not experience prolonged waiting periods. This aspect significantly influences patient retention and the likelihood of revisiting the same medical facility or doctor. By incorporating this question into the patient satisfaction survey, valuable insights can be gathered regarding the efficiency of appointment scheduling, enabling improvements in the process to enhance overall patient contentment.

Q3: What is the assessment of the professionalism demonstrated by our staff?

This should answer how would the venture evaluate the demeanour and professionalism of the healthcare providers, including nurses and doctors, as well as the administrative and support staff using a better digital system (Figure 5.1).

Q4: How would you assess the diagnostic process you underwent during the investigation phase?

Patients frequently arrive at medical facilities experiencing localised pain and may struggle to articulate their symptoms to healthcare professionals. The investigative diagnostic process involves a comprehensive approach, considering patient history, family history, conducting necessary tests, and administering essential medication.

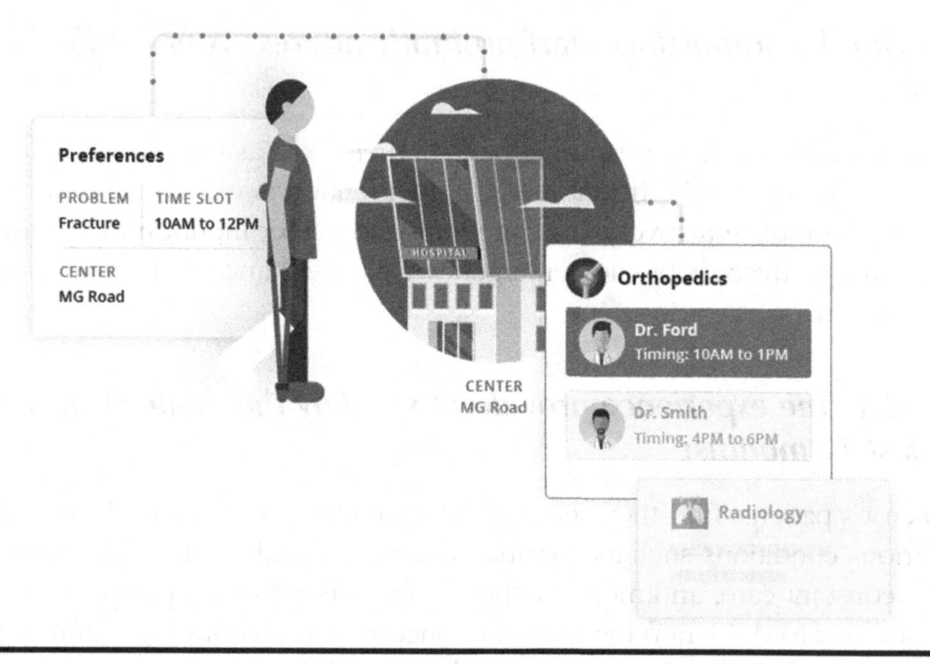

Figure 5.1 Demonstration for a better digital system incorporation [8].

Q5: How often did you receive conflicting information from different medical care professionals at this hospital?

Conflicting information from various healthcare professionals can lead to confusion and negatively impact the patient experience. This question aims to gauge the consistency of information provided by different medical care professionals within the hospital. Analysing instances of conflicting information from different medical care professionals is a crucial measure to assess the competency of the current staff. This evaluation allows for the identification of areas that may require process improvement, ensuring consistent and accurate information dissemination to patients.

Q6: What is the difference in the care provided by the hospitals available in your area?

Medical facilities should consistently deliver top-notch healthcare services to all patients. Furthermore, it is imperative to leverage internal strengths and incorporate effective practices observed in neighbouring hospitals.

Q7: Did the ambulatory staff promptly address your medical care needs?

Rapid-response medical personnel play a crucial role as primary healthcare providers. Prompt and efficient medical intervention is vital for patients to achieve a complete recovery. It is imperative to gather input on the promptness and effectiveness of each emergency call to enhance services and overall efficiency.

Q8: Rate the experience provided to you by the medical staff in the last 12 months?

Numerous patients find themselves in an extended medical care facility due to various conditions such as specific ailments, mental health issues necessitating constant care, and more. Gathering insights into their perspectives on the care received, identifying areas of concern, and recognising commendable aspects contributes to enhancing the overall care provision experience. Whether the staff showed pity, empathy, sympathy, or compassion should be answered (Figure 5.2).

An effective patient satisfaction survey should encompass well-crafted questions designed to maximise response rates. These questions should provide a comprehensive overview of a patient's medical care experience, ensuring simplicity in response while incorporating a combination

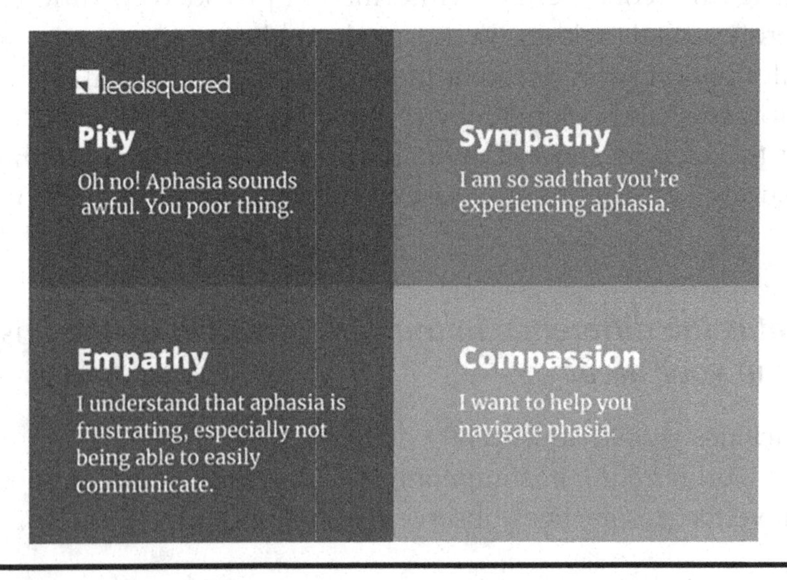

Figure 5.2 Staff behaviour [8].

of open-ended and other question types. Additionally, utilising the survey results to identify problem areas and gain insights from patients helps gauge the severity of issues and identifies areas for improvement.

In numerous instances, specific aspects of a medical facility's operations may go unexamined, attributed to the scarcity of patient feedback and a general lack of information about the challenges faced by patients during consultations and treatments [9]. Patient satisfaction surveys serve as a valuable platform for patients to offer candid feedback based on their experiences. Let's explore the significance of conducting regular patient satisfaction surveys and delve into the following compelling reasons for their implementation in maintaining and expanding patient engagement.

- Charting and monitoring a patient's journey: It is essential for a healthcare institution to monitor patient satisfaction concerning the care they receive. To guarantee optimal patient care at each crucial interaction point with the medical staff and the institution as a whole, conducting patient satisfaction surveys at various stages of their journey is highly advisable. This practice facilitates the identification of areas with low satisfaction levels, allowing for appropriate interventions.
- Document instances of staff misconduct: The use of a patient satisfaction survey template encourages transparency, offering patients a platform to report any mistreatment by hospital staff and bring it to the attention of the management. The management can then take appropriate action, including termination or training, to ensure that all staff members treat every patient with respect during their visit to the medical facility.
- Enhancing patient care: Patient satisfaction surveys enable medical institutions to assess the disparity between patient expectations and the services provided. This process enhances visibility into operational efficiency, allowing medical facilities to address identified gaps. In an era where every industry is becoming customer-centric, the healthcare sector is no exception. Numerous aspects of a medical institution's operations may require improvement, making direct feedback from patients invaluable. Including questions in the survey about ways to enhance patient service facilitates continuous evolution in alignment with shifting patient perspectives.
- Provide prompt services to patients: While 80% of medical professionals believed they consistently adhered to patients' appointment schedules, nearly 50% of patients held a different view, indicating that they were not called in at their designated appointment times.

- Obtain insights on hygiene standards: The cleanliness and hygiene of a hospital serve as indicators of its overall maintenance. A considerable number of patients prioritise visiting medical facilities with clean and hygienic environments. Including questions related to this aspect in the survey questionnaire can provide valuable insights.

5.2 Understanding Market Competition

Attaining an in-depth understanding of the strengths and weaknesses of competitors, along with identifying factors resonating with the shared audience, is crucial for enhancing various facets of marketing strategy, encompassing branding, search engine optimisation (SEO), paid search, and social media. Recognising effective and ineffective marketing tactics is just the starting point; a thorough competitive analysis delves deeper, pinpointing potential gaps in marketing efforts. This insight facilitates the formulation of innovative and compelling strategies to fortify the approach, attract a broader patient base, and foster business growth. In the highly competitive healthcare industry, achieving high rankings on search engines necessitates meticulous evaluation and comparison of marketing initiatives against competitors. Comprehensive understanding of aspirational, direct, and secondary competitors is imperative for informed decision-making and achieving success [10].

Conducting a competitive analysis is essential for identifying key competitors in the industry and examining their marketing strategies [11, 12]. This analysis takes precedence in the overall branding and digital marketing strategy, especially during the launch of a new brand, a rebranding effort, or the consolidation of multiple brands. However, to execute a well-informed competitive analysis, it is crucial to first define the brand's position in the marketplace [13]. A research-driven positioning statement plays a pivotal role in communicating the brand's unique value to customers in relation to primary competitors. This positioning statement typically encompasses four essential components [10]:

- Target audience: Illustrate the demographics, geographical locations, values, and attitudes that the brand aims to engage and attract. These elements play a significant role in shaping the content, branding, and creative components.
- Market definition: Establish the category in which the brand competes and its relevance to the customers. Identify the keywords utilised by the

target audience, even if they differ from the preferred terms. This information will significantly contribute to the SEO analysis.

- Brand promise: Identify the most compelling benefit, whether emotional or rational, that a brand can uniquely offer to the target audience in comparison to the competition. Gaining insights into what resonates with the target audience will guide the analysis of organic, local, and paid search strategies.
- Evidence of credibility: Clearly articulate why customers should have faith in the brand and how it fulfils its commitments. Elements such as demonstrated outcomes and endorsements will be considered in the analysis of social media.

Numerous hospitals and providers with multiple locations may have already crafted a robust brand positioning statement. Nevertheless, it is essential to routinely reassess and refine it to ensure that the brand promise and the "reason to believe" remain current and are tailored to meet the dynamic needs of the increasingly informed healthcare consumer.

Commencing a comprehensive competitive analysis involves the compilation of a roster featuring aspirational, primary, and secondary competitors, with a particular emphasis on formidable entities. Secondary competitors encompass healthcare organisations that engage in indirect competition, such as those situated outside catchment areas or specialising in niche treatment modalities, including services like telemedicine. It is imperative to accord due consideration to emerging healthcare entities [10].

Upon finalising the list, the subsequent step involves a meticulous examination of each competitor's array of services and a thorough comprehension of their marketing strategies targeted at patients. Gaining insight into how patients and referring physicians perceive competitors provides valuable information regarding whether hospitals or multi-location practices deliver equivalent or superior quality and service offerings [14]. Conducting a comprehensive competitive analysis within the healthcare sector also facilitates the identification of distinctive values that competitors either lack or cannot furnish. These unique values can then be strategically highlighted in marketing approaches.

Unlike the other marketing branches discussed in this chapter, branding is highly subjective. Nevertheless, marketers and other creatives must be prepared to support design decisions with relevant data and research [15].

Conducting a brand and creative competitive analysis aids businesses in refining (or establishing) content that effectively converts by leveraging insights from the brand positioning statement. It provides the rationale necessary to underpin creative choices. Whether the goal is to rejuvenate brand and creative aspects or explore partnerships with other brands, three distinct areas warrant attention during the competitive analysis [10]:

- Brand audit: Conducting a brand audit represents a crucial initial step in enhancing the identity, as it elucidates the current alignment of the brand and creative elements, identifies areas where it may have deviated, and discerns preferences regarding specific elements. This phase also aids in visualising the reasons behind varying levels of engagement in specific channels.
- Competitor audit: A competitor audit sheds light on how key players position themselves in the marketplace, addressing questions such as, "Which colours and typefaces are employed by brands?" "Is there a consistent theme in their logos utilising symbols or other iconography?" "Does their tone lean towards formality, casualness, seriousness, or humour?" Based on the insights derived, considering an alternative approach may be warranted. This exercise is designed to pinpoint creative gaps and formulate compelling strategies to address them, ultimately broadening the appeal to a wider audience.
- Demographic audit: A demographic audit is indispensable for comprehending the target audience and brand personas. Is the designated age group appropriate for the targeting efforts? Are considerations given to local audiences? This audit ensures that businesses convey the appropriate narrative to the right individuals at each stage of the customer journey. Additionally, it aids brands in identifying potential messaging opportunities and discerning content gaps.

Gaining competitive insights for paid social media is crucial to discern the whereabouts of the audience and what resonates with them. A social media competitive analysis provides valuable information to hospitals and multilocation providers regarding the strategies employed by key competitors in engaging with their audience. Typically, a social media competitive analysis is conducted prior to the initiation of a new campaign or the optimisation of an existing one.

Social media is perpetually evolving, yet a constant remains: it provides businesses the opportunity to connect with and engage potential customers on platforms such as Facebook, Instagram, LinkedIn, Twitter, YouTube, and more. For hospitals and health systems, a robust social media strategy is instrumental in audience engagement and brand growth.

In the current landscape, key competitors are likely actively interacting with the target audience. Therefore, aligning the social media strategy with theirs becomes imperative. However, this alignment should not mark the conclusion of efforts. To make a meaningful impact, it is essential to discern their actions, identify areas where they may be lacking, pinpoint the platforms they utilise, and strategise on how to enhance the approach. Competitive analysis in social media marketing centres around three key focus areas [10]:

- Channels: A variety of social media channels are available to choose from; however, understanding the primary platforms where both the organisation and competitors are active is crucial. Identifying the channels that yield the highest engagement is equally important. In the healthcare sector, organisations often discover that their audiences prefer platforms such as Facebook, Twitter, and LinkedIn. If it is observed that a competitor lacks a substantial presence on Twitter, there may be an opportunity to enhance visibility by emphasising value-based messaging on this channel, thereby attracting a new audience.
- Paid promotional content: Do competitors deploy paid advertisements across their social media channels? If affirmative, comprehending the nature of these advertisements, the language employed in their copy, the incentives they feature, and the calls to action (CTAs) they include is crucial. This knowledge assists in determining whether their primary goal is to elevate brand recognition, generate leads, drive conversions, etc. Understanding these objectives and scrutinising their paid promotional content yields valuable insights into their overarching paid social strategy.
- Unpaid or organic content: While many healthcare organisations utilise paid content for swift audience expansion, it remains crucial to continually build a strong foundation of organic content for sustained support. Unpaid content provides an opportunity to showcase creative expertise and engage the audience in more unique ways. The primary goal of organic content is to deliver value to the audience that sets it apart from competitors' offerings.

  Investing time in gaining a deep understanding of aspirational, direct, and secondary competitors proves crucial for making informed decisions. This knowledge, in turn, increases the likelihood of enhancing Search Engine Results Page (SERP) ranking, attracting a larger and more engaged audience, and fostering business growth.

In conclusion, it is noteworthy that a competitive analysis within the healthcare industry for any marketing branch is subject to change over time. Regularly conducting these analyses is imperative to remain competitive, dominate in organic and paid search results, and excel in local search rankings.

References

1. Zgierska A, Rabago D, Miller MM. Impact of patient satisfaction ratings on physicians and clinical care. *Patient Prefer Adherence* 2014;8:437–46. https://doi.org/10.2147/PPA.S59077.
2. Customer satisfaction surveys: a comprehensive guide n.d. https://www.helpscout.com/blog/customer-survey/ (accessed January 7, 2024).
3. Patient satisfaction surveys: building loyalty and delivering better care n.d. https://www.zonkafeedback.com/blog/patient-satisfaction-surveys (accessed January 7, 2024).
4. Khuangsirikul S, Lekkreusuwan K, Chotanaphuti T. 10-Year patient satisfaction compared between computer-assisted navigation and conventional techniques in minimally invasive surgery total knee arthroplasty. *Comput Assist Surg* 2016;21:172–5. https://doi.org/10.1080/24699322.2016.1249959.
5. Freeman T, Jolley G, Baum F, Lawless A, Javanparast S, Labonté R. Community assessment workshops: a group method for gathering client experiences of health services. *Health Soc Care Community* 2014;22:47–56. https://doi.org/10.1111/HSC.12060.
6. 20 Patient satisfaction survey questions for questionnaire n.d. https://www.questionpro.com/blog/patient-satisfaction-survey/ (accessed January 7, 2024).
7. Buers C, Triemstra M, Bloemendal E, Zwijnenberg NC, Hendriks M, Delnoij DMJ. The value of cognitive interviewing for optimizing a patient experience survey. *Int J Soc Res Methodol* 2014;17:325–40. https://doi.org/10.1080/13645579.2012.750830.
8. 10 Best patient satisfaction survey questions - LeadSquared n.d. https://www.leadsquared.com/industries/healthcare/patient-satisfaction-survey-questions/ (accessed January 7, 2024).
9. Kazzazi F, Haggie R, Forouhi P, Kazzazi N, Malata CM. Utilizing the total design method in medicine: maximizing response rates in long, non-incentivized, personal questionnaire postal surveys. *Patient Relat Outcome Meas* 2018;9:169–72. https://doi.org/10.2147/PROM.S156109.
10. Healthcare industry competitive analysis n.d. https://healthcaresuccess.com/blog/healthcare-marketing/your-guide-to-healthcare-industry-competitive-analysis-for-branding-digital-marketing.html (accessed January 7, 2024).
11. Alibrahim A, Wu S. Modelling competition in health care markets as a complex adaptive system: an agent-based framework. *Heal Syst* 2020;9:212–25. https://doi.org/10.1080/20476965.2019.1569480.

12. De Silva DG, Jung H, Kosmopoulou G. The impact of regional competition on the health care industry. *Appl Econ* 2018;50:5135–41. https://doi.org/10.1080/00036846.2018.1467551.
13. Awoyemi BO, Olaniyan O. The effects of market concentration on health care price and quality in hospital markets in Ibadan, Nigeria. *J Mark Access Heal Policy* 2021;9. https://doi.org/10.1080/20016689.2021.1938895.
14. Jiang Q, Tian F, Liu Z, Pan J. Hospital competition and unplanned readmission: evidence from a systematic review. *Risk Manag Healthc Policy* 2021;14:473–89. https://doi.org/10.2147/RMHP.S290643.
15. Simonet D, Katsos JE. Market reforms in the French healthcare system: between regulation and yardstick competition. *Public Money Manag* 2022;42:191–8. https://doi.org/10.1080/09540962.2020.1752467.

DESIGN

Chapter 6

Design Thinking and Product Design Process

6.1 Understanding Why New Products Fail

When an entrepreneur launches a new product or service, it can be complex and costly to make products a success in the market [1]. These complexities could lead to product failures. These failures could not only be technical malfunctioning but could include minimal sales due to lack of product-market fit (PMF) [2]. A general term used for these failures in the entrepreneurship is "product failure." Product failure, in summary, refers to a product that failed to achieve the intended results, be it of a financial nature, user acceptance, or a comparable kind. Failing to achieve the intended results, which are frequently assessed using strategic objectives and key results metrics, constitutes failure [3]. Goals may encompass various dimensions and extend beyond mere financial targets of revenue and profit. For instance, they may involve creating a novel upsell opportunity for another product or establishing a fresh competitive advantage. Figure 6.1 shows a typical product-market fit pyramid. It shows how the product having correct user experience (UX), feature set, and value proposition should be well connected to market's undeserved needs and should reach their target customer.

Popular data estimated that 80–95% of novel products ultimately fail [5]. Although more recent studies have the figure closer to 40%, the probability of product failure continues to pose a significant and tangible threat to nascent enterprises. It is not a simple task for new ventures to introduce

DOI: 10.4324/9781003475309-9

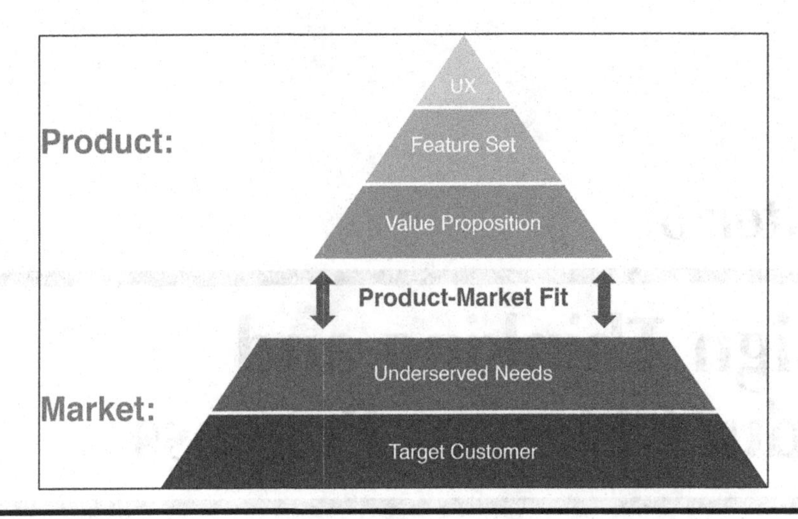

Figure 6.1 An illustration representing the product-market fit (PMF) pyramid [4].

items that generate sufficient cash to support the operation, despite the fact that the risks are enormous. Hence it is important to study the reasons, and past examples which experienced the overall product failure.

One of the reasons for product failure is the lack of PMF. Without PMF, it is improbable that a product would achieve significant success, whatever its efficacy [6]. In 2006, Microsoft resolved to compete with the iPod. The firm introduced Zune, a device that claimed to do all the functions of the Apple iPhone as well. Yet, despite its lofty potential, Zune was a commercial failure. Pursuing Apple, Microsoft developed a product that failed to provide any justifications for users to transfer. In comparison to Apple, Microsoft's marketing strategy was ineffective, and the Zune's feature set failed to provide customers with the same level of value as the iPod's. Mistargeting the market or pursuing a market that is too small to support the firm might result in fit issues [7, 8]. Businesses that consistently achieve and maintain PMF do so by adjusting their products to meet the evolving demands of their target market. In another example, the Segway personal mobility gadget failed because it failed to address a genuine market concern. It was an innovation in technology and not in consumer value. Figure 6.2 represents the process of PMF. After performing product iterations, the viability, desirability, and feasibility together pave the way for the PMF and lead to further scaling of the product.

Another reason could be that the new technology requires too much understanding and education [10]. Many emerging businesses aspire to distinguish themselves from the competition through the establishment of a novel market segment. Regrettably, the general public may comprehend more than the mere definition of a new category in order to comprehend

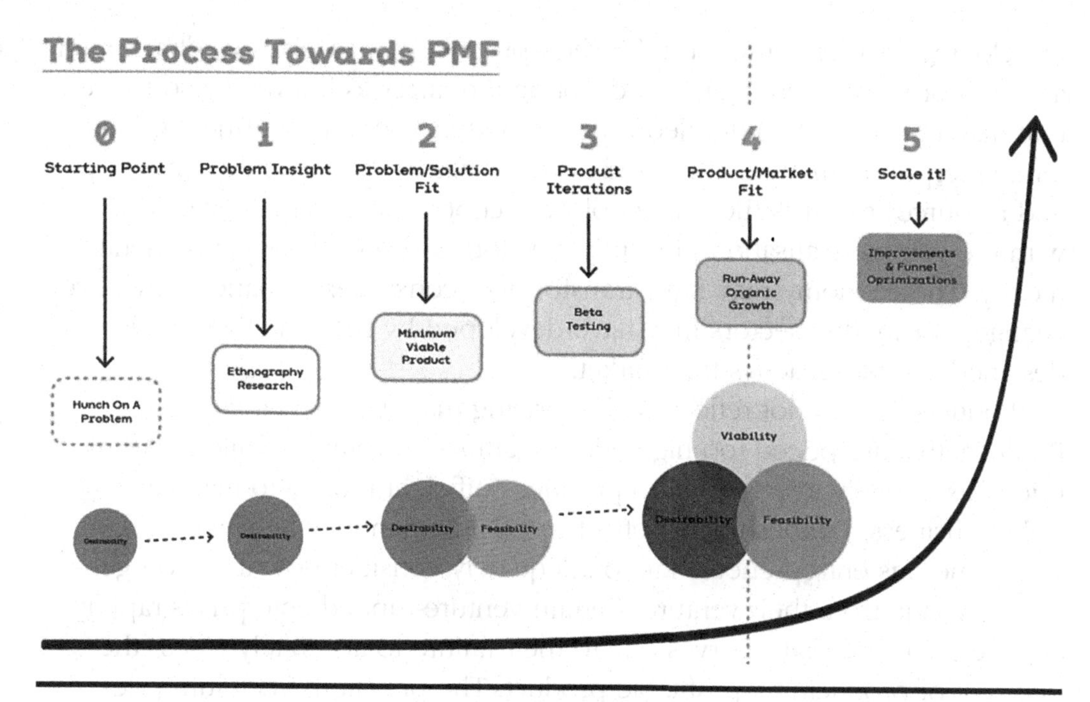

Figure 6.2 The possible process to reach PMF [9].

how a product is distinctive and will address their difficulties. It is imperative that individuals are adequately informed and persuaded that a product or service addresses a distinct void in the market. Furthermore, it is important that buyers perceive the solution as a genuinely exceptional value addition, surpassing which the current competitors are incapable. In the event that education materials are inadequate in quantity or fail to present a persuasive enough argument in favour of category formation, customers will persist in their loyalty towards established brands.

Weak product launches could also result in early product failures [11]. A company's product release is an eventful occasion that merits publicity. However, mere commemoration of the launch is insufficient. Conversely, a substantial amount of work must be devoted to brand recognition and advertising in order to generate support for the product [12]. Product failure is inevitable when a restricted budget and promotional effort are combined with a lack of social media engagement and consumer support. The absence of expanding brand recognition and overall product marketing will result in a lack of demand for products [13].

When introducing anything new to the market, an entrepreneur may strike a balance between the value they want to provide and the amount of time their team needs to produce it. A delayed product launch is just as problematic as introducing a product that is just partially developed.

Developing and releasing an unfinished product that receives feedback after a month of work is preferable to devoting a quarter to improving software that may not solve the intended problem. Often, the failure of new goods (and enterprises) may be attributed to the excessive allocation of time and resources towards the pursuit of perfection, rather than proceeding with the commercialisation of a fully developed product that can generate money. The economy may have transformed, consumer demands may have changed, or market sectors may have developed by the time that overly designed product reaches the market.

Products that do not reflect market pricing may experience failure [10]. Products that are priced too high will not attract customers, while those priced excessively low will fail to produce sufficient money to maintain a viable business. Establishing an effective pricing plan may be challenging, and numerous entrepreneurs fail to adequately consider how they will generate revenue from their venture. Certain venture-funded enterprises rapidly deplete their financial reserves due to their failure to accurately assess the true cost of producing a profitable product. The prevalent freemium pricing model confronts companies with an additional obstacle. It might be more challenging to convince clients to upgrade from the free version of a software to the premium flagship product by requiring them to open their wallets. Users will not feel compelled to upgrade to the premium edition of a programme if the free version provides sufficient perceived value. In the meanwhile, while waiting for free users to convert into paying customers, a new company may exhaust its resources. If the price strategy is impeding the success of a product, it is necessary to make adjustments. Additionally, consider the true cost of producing the goods. Figure 6.3 shows an early-stage wearable personal computer (Xybernaut Poma Wearable PC, Xybernaut and Hitachi) which was priced at around $1,500. Due to the price tag and poor wearability, the product was a failure and was withdrawn from the market.

All reasons apart, if the product design and quality is not at par with the products or service in the market, the chances of failure are higher. It is improbable that a product of inferior quality or design would succeed in attracting and retaining customers. The Google Glass wearable device, for instance, was unsuccessful due to its weight, expense, and lack of practical functionality. Additionally, it prompted privacy issues and garnered derision from its users [15]. Defective design may lead to the attrition of a devoted clientele, negative evaluations, and significant erosion of profit margins. Figure 6.4 shows an example of a bad product design where the "cat-like" protrusion disturbs while drinking anything from it.

Figure 6.3 An early-stage wearable personal computer launched in 2002 by Xybernaut and Hitachi [14].

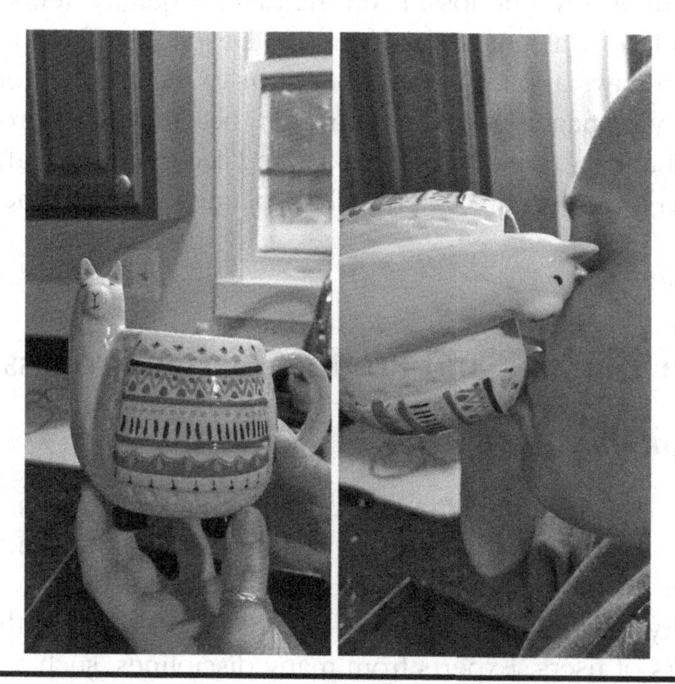

Figure 6.4 An example of bad product design [16].

Figure 6.5 An example of poorly designed website [17].

The same holds true for websites that are poorly designed. Figure 6.5 shows an example of a poorly designed website. Often, badly designed websites mislead visitors into believing that the items are similarly unusable, which can result in revenue loss. Even the highest quality items might be adversely affected by a substandard website. Despite the fact that a number of these design flaws appear to be rather clear to the corporation that created them, they managed to make it through the production process without being deemed abject. Hence, in the upcoming section, design thinking and product design processes are extensively explained to overcome product failures.

6.2 Design Thinking and Product Design Process

6.2.1 Design Thinking

Teams employ design thinking, a non-linear, iterative approach, to comprehend consumers, reframe challenges, question assumptions, and generate novel ideas for prototyping and testing [18, 19]. It is essential to cultivate and hone the ability to comprehend and respond to swift shifts in the settings and behaviours of users. Experts from many disciplines, such as engineering and architecture, later refined this extraordinarily innovative procedure in order to tackle contemporary human requirements. Design thinking enables design teams to address unclear or ill-defined challenges by reorienting them in a user-centric manner and prioritising what is most crucial [20–23]. Design thinking is very definitely the most effective of all design approaches for "outside-the-box thinking." Through improved usability testing, prototyping, and UX research, teams may discover novel approaches to satisfying users' requirements.

Design thinking is a scientific and artistic endeavour. It integrates inquiries into uncertain aspects of the issue with logical and analytical investigation—or, to put it another way, the scientific aspect [24]. Engaging in unconventional thinking can yield a novel resolution to a challenging dilemma. However, the ability to think creatively may be rather difficult to achieve due to the fact that we create thought patterns spontaneously, which are influenced by the repeated tasks and readily available information that we encounter. The Design Thinking methodology is extremely user-centric and forward-thinking. Let's begin by examining the procedure in further depth in light of the four Design Thinking principles established by Harry Leifer and Christoph Meinel of the Hasso-Plattner-Institute of Design at Stanford University, California.

Principle 1 is titled as "the human rule." All design work is social in nature, irrespective of context, and any social innovation will force us to return to a "human-centric perspective." Principle 2 is described as "the ambiguity rule." Ambiguity is unavoidable and resistant to elimination or oversimplification. It is essential to conduct experiments at the boundaries of knowledge and capability in order to gain a fresh perspective. Principle 3 describes the "redesign rule" in which societal conditions and technological advancements may shift, but the fundamental human needs remain constant. The last, Principle 4, is titled as "the tangibility rule." It suggests the implementation of prototypes allows designers to convey concepts with more efficacy.

In addition to this, there are five design thinking phases that an entrepreneur can consider—Empathise, Define, Ideate, Prototype, and Test—are most effective when used to unknown or ill-defined situations [25]. Figure 6.6 shows the schematic for design thinking.

6.2.1.1 Phase 1: Empathise

In the initial phase of design thinking, user-centric research is prioritised. An entrepreneur may desire to get a sympathetic comprehension of the issue they are endeavouring to resolve. Conduct observations to engage and

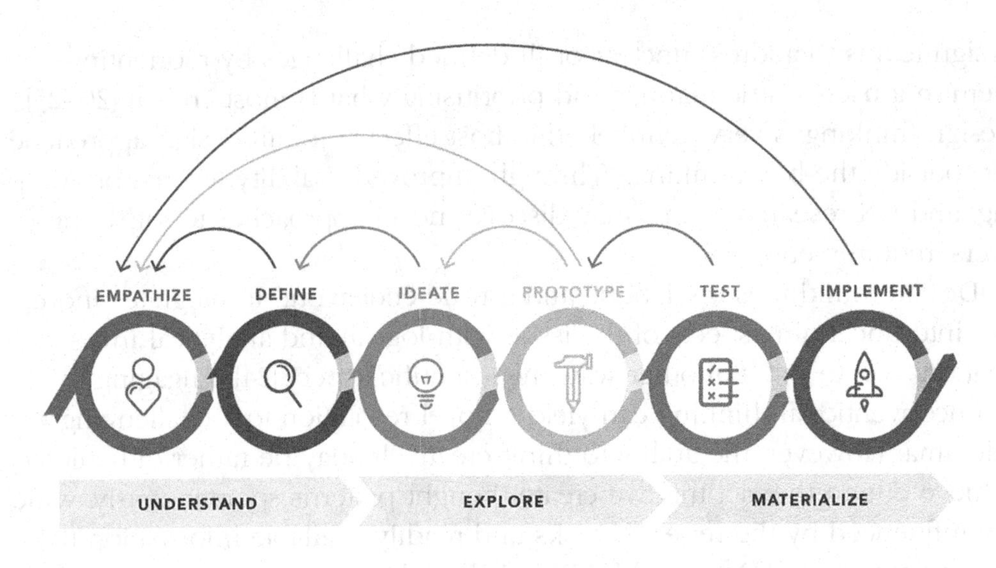

Figure 6.6 Schematic for design thinking [26].

sympathise with their users and consult specialists for extra information on the area of issue. In addition to their experiences and motives, an entrepreneur may choose to obtain a more profound, personal knowledge of the issues at hand by immersing themselves in the physical world of their consumers. Empathy is fundamental to the human-centred design process and issue resolution, as it enables design thinkers to abandon their preconceived notions about the world and acquire genuine understanding of users and their requirements. An entrepreneur may amass a considerable volume of data for utilisation in the subsequent phase, contingent upon temporal limitations. The primary objective of the Empathise phase is to gain the most comprehensive knowledge possible of consumers, their challenges, and the requirements that form the basis for the product or service an entrepreneur want to produce.

6.2.1.2 Phase 2: Define

Determining the problem constitutes the second phase of the Design Thinking algorithm. An entrepreneur may compile and begin to make sense of all the facts gathered throughout the Empathise phase: what challenges and obstacles are their users encountering? What patterns are evident to an entrepreneur? What is the most significant user issue that their team must address? Upon completion of the Define step, a well-defined problem statement will be obtained. Ultimately, the objective is to surmount cognitive fixedness and generate novel and inventive concepts that address the issues that have been recognised. The Define phase will facilitate the collection of brilliant concepts by the design team for the purpose of establishing features, functionalities, and other pieces that either address the problem at hand or, failing that, enable actual users to resolve difficulties with little effort.

6.2.1.3 Phase 3: Ideate

Commencing with a well-defined issue statement and a comprehensive understanding of the consumers, the next step is to generate alternative solutions. Creativity occurs during the ideation stage, which is the third phase of the Design Thinking process. It is essential to note that at this stage, no judgement is permitted. To generate a wider range of novel perspectives and concepts, designers will conduct ideation sessions. Designers may employ a variety of ideation techniques, including mindmapping and brainstorming, bodystorming (roleplay scenarios), and provocation—an extreme kind of lateral thinking that encourages the designer to question preconceived notions and investigate new alternatives and solutions. By the conclusion of the ideation phase, a limited number of concepts will remain for further development.

6.2.1.4 Phase 4: Prototyping

To test the primary ideas created during the ideation phase, the design team will now construct a number of affordable, scaled-down prototypes of the product (or particular features found inside the product). The prototypes may be disseminated and evaluated by members of the design team, other departments, or a select subset of individuals. During this experimental phase, the optimal solution for each of the issues identified in the preceding three phases must be determined. The answers are implemented within the prototypes, and subsequently assessed, refined, or rejected in accordance with the users' experiences, one at a time. As the prototype phase concludes, the design team may possess a more comprehensive understanding of the product's constraints and the challenges it encounters. In addition, they will have a more precise understanding of how authentic people will act, think, and feel while interacting with the final product.

6.2.1.5 Phase 5: Test

Through thorough testing, designers or evaluators apply the most effective solutions uncovered in the prototype stage to the final product. While this concludes the five-stage paradigm, in an iterative process like design thinking, the outcomes produced are frequently utilised to reframe one or more additional challenges. This heightened degree of comprehension might potentially facilitate an examination of the product's usage conditions and individuals' thoughts, actions, and emotions around it. Moreover, it might prompt a return to an earlier phase in the design thinking procedure. Then, more cycles might be conducted in which modifications and adjustments are implemented in order to exclude alternate alternatives. Gaining a comprehensive grasp of the product and its users is the ultimate objective.

6.2.2 Product Design Process

Product design encompasses many key stages: market opportunity identification, problem definition with precision, solution development, and user validation with actual users [27]. A well-known paradigm known as Design Thinking, as explained in the previous section, underpins product design. It is an innovative method that prioritises human needs, technology capabilities, and commercial objectives [28, 29]. A Product Design Process comprises a series of deliberate activities designed to transform a concept into a commercially viable product for the end customer.

Contemporary technologies constitute an extraordinarily intricate social enterprise. Therefore, research and development, design, management, marketing, financing, manufacturing, production, and maintenance are all components of the product design process. Due to its complexity, engineering product designs cannot be managed independently by a single individual. Given the magnitude of stakeholder engagement, it is imperative to establish well-structured procedures and maintain efficient communication among all parties involved. Therefore, in order to enhance the likelihood of a successful engineering product design, it is imperative to strategise and implement a methodical design approach.

The design process may differ based on the nature of the product (e.g., whether it is tangible or immaterial). Illustratively, tangible items include smart watches, health trackers, non-invasive glucometers, etc. which need meticulous design and manufacturing throughout the product design process [30, 31]. Conversely, implementation, testing, verification, and deployment will occur for intangible items such as computer software, services, and mobile phone apps, as opposed to detailed design and production. Product design procedures may also undergo modifications contingent upon the driving force behind engineering product innovation—whether it be invention-push (technology-push) or demand-pull (market-pull) innovation.

In the absence of a product design process plan, the intricacy of the process and the multitude of accessible methodologies would provide product designers with an insurmountable array of alternatives. Therefore, designers are required to acquire an understanding of the design process, diverse methodologies, and the operational and decision-making procedures outlined in procedural blueprints. The product design process could be divided into four following phases [32].

6.2.2.1 Phase 1: Product Definition and Planning (PDP)

Despite being the most crucial phase in the product design process, product definition is frequently disregarded by organisations and product designers. "Needs analysis," "Problem definition," "Product planning," and "Task clarification" are other terms that may be applied, contingent upon the business, product, and new product development stage. Product planning is often conducted by marketing, management, and product managers at the earliest phases of the PDP or new product development. Due to budgetary considerations, many PDP decisions are unfortunately finalised prior to the engineering design phase. This is not ideal for the product designers. Task clarification occurs when engineering assumes responsibility for a client's specification, which may be an incomplete product definition. In order to ascertain how customers view the products, services, and brand of an organisation, it is necessary to perform a customer requirements analysis.

The primary aim of the product definition activity is to acquire a comprehensive understanding of the product need or problem. Regardless of the nature of the product being developed, it is imperative that the design team commences by ascertaining the technical specifications of the product through an analysis of the client's requirements and the subsequent creation of a product design specification (PDS). The design and development team normally receives the product design request in one of the following methods. A written or verbal request accompanied by a problem statement from an external client or internal department could be one method. A concise synopsis of fundamental product prerequisite could be another method. In the event that the need is innovation or technology-driven, the design team is obligated to provide product concepts that include the novel technology could be the last choice depending on the scenario.

The product definition phase therefore facilitates the following for the design team. First, it guarantees that the engineers and designers thoroughly analyse and comprehend the product's requirements. Second, a written formal document, known as a PDS, is generated as a means of verifying and evaluating the product. Third, it aids the customer in providing more specific explanations so that the designers may comprehend the nuances.

6.2.2.2 *Phase 2: Conceptual Design*

In the early phase of the design process known as Conceptual Design, the fundamental functions and aesthetics of a given object are defined. It encompasses the development of strategies, processes, relationships, and experiences. Early on in the engineering product design process, concept or conceptual design is an element of product design. Typically positioned between the product definition and embodiment design is the new product development cycle.

Demand-pull product innovations may involve the early elaboration of product requirements, while the product design process encompasses the concept generating phase. However, the front end of technology-push new product creation is very perplexing, given that while the technology itself is established, the precise manner in which it may be applied to build commercially viable goods is not entirely clear. Therefore, idea design is where an entrepreneur begins, with a fuzzy finish where they do experiments. In the nebulous realm of new product development, the identification of the product's range, design, and other aspects is significantly influenced by idea and technology development. The focus of the idea or conceptual design stage is the design that was selected during the product design phase.

Different stages of concept design include clarifying the problem, searching the problems externally and then internally, exploring systematically, reflecting, and choosing the required concept. Clarifying the problem further includes understanding the user and their requirements, defining the problem statement from their perspective, identifying the constraints in progressing the solution, breaking down the problem into small segments, and simultaneously prioritising the small problems identified in the previous step. Searching externally and then internally could include literature searches, patent searches, review articles, competition analysis, expert guidelines, attending events, talking to peers, and reviewing of existing design and product specifications. At the end, determine the design factors, define the value range for each variable in the design, produce every conceivable permutation of the design variables, assess every possible permutation of design factors, determine the most auspicious permutations of design factors to be investigated further, rank and prioritise the concepts, and get feedback from others.

One primary benefit of conceptual design is that it elucidates the capabilities and intended purpose of a product. In the process of product development, this is vital. Conceptual designs facilitate the development of intuitive and user-friendly user interfaces. Detailed descriptions of the roles and responsibilities of the various project users contribute to a greater comprehension of the undertaking. The design may facilitate thoughtful consideration of the user's perspective, so simplifying the process of attaining the intended outcomes.

6.2.2.3 Phase 3: Embodiment Design

One of the primary phases of the product design process, embodiment design involves the development of the principal engineering product design concept in accordance with the product design specification and economic criteria. This development reaches a level where further detailed design can proceed directly into production. Due to the fact that each technical product design is unique and contributes to the aforementioned difficulties, it is exceedingly difficult to establish rigid blueprints for the embodiment design phase. Consequently, during the embodiment phase of new product development, a functional system or product will be fashioned from an abstract design concept that remains within the allotted unit cost.

Under this phase, three sub-phases are required, i.e., product architecture, design configuration, and parametric design. System-level design, which is another name for product architecture, consists of delineating and assigning physical entities or components to their respective functions inside a product. Physical components, sometimes referred to as modules, are described and organised in a manner that fulfils the overarching product requirement specification.

Modular design involves the subdivision of the overarching product purpose or system-level function into more manageable single functions or specific processes. These are then assigned to discrete components or subassemblies known as modules. These modules are regarded as discrete constituents and possess well-delineated mechanical or electrical connections.

Once these components are interfaced, the product as a whole is formed and able to fulfil its whole purpose. The prevalence of engineering products with modular architectures surpasses that of integrated designs. When economic, environmental, and weight restrictions compromise the performance of a product, an integrated system-level design is frequently favoured. Design for production and assembly, which prioritises the reduction of components, is an additional significant factor that influences the development of integrated product architecture products. In general, integral engineering products have an exceptionally high function-to-components ratio, which indicates that the same components may be utilised in a variety of ways to fulfil various purposes.

In design configuration, shape and general dimensions or sizes are established for the components defined in product architecture. It is mainly dependent on the three-dimensional constraints that define the envelope in which the product operates and the product architecture. This would be a preliminary selection of material, manufacturing process, modelling, sizing of parts, etc.; on the other hand, the primary aim of parametric design is to assign values to design variables in order to generate an optimal product design or functional component, taking into account economic and technological considerations.

The analytical nature of this design element surpasses that of conceptual or design configuration. The design variable refers to a characteristic of a component that is determined by the designer. Typical design variables include but are not limited to dimension, tolerance, material, surface finish, and heat treatment. The primary aim of parametric design is to determine the optimal values for the design variables in order to generate the most cost-effective and performance-optimal design feasible. In addition to determining the dimensions and tolerances, parametric design aims to optimise performance and quality while reducing expenses.

6.2.2.4 Phase 4: Detailed Design

The designing team finalises the embodiment design and generates a comprehensive specification for every component and assembly during the Detailed Design phase. This consists of the final specifications and particulars on the shapes, forms, functions, dimensions, material possibilities, and surface finishes of the parts. The stages and procedures of the detailed design vary according to the product type, the method of developing new products, and the organisational structure of the firm. Typically, detailed design steps might comprise the following steps. First, finalise original equipment manufacturer component specs and detailed drawings. Second, organise assemblies by integrating separate components and generating bills of materials and drawings. Third, incorporate manufacturing notes, assembly procedures, transport, and operation instructions, and more into the final production papers. Fourth, verify the standards, completeness, and accuracy of each document.

References

1. Chang CT, Chen PC, (Marcos) Chu XY, Kung MT, Huang YF. Is cash always king? Bundling product–cause fit and product type in cause-related marketing. *Psychol Mark* 2018;35:990–1009. https://doi.org/10.1002/MAR.21151.
2. Olsen D. Achieving product-market fit with the lean product process. *Lean Prod Playb* 2015:1–12. https://doi.org/10.1002/9781119154822.CH1.
3. 7 Reasons why products fail | 280 Group n.d. https://280group.com/product-management-blog/7-reasons-products-fail/ (accessed November 28, 2023).
4. Program LSCE. A playbook for achieving product-market fit the lean way n.d. https://leanstartup.co/resources/articles/a-playbook-for-achieving-product-market-fit/ (accessed November 28, 2023).
5. Product innovation: 95% of new products miss the mark | MIT Professional Education n.d. https://professionalprograms.mit.edu/blog/design/why-95-of-new-products-miss-the-mark-and-how-yours-can-avoid-the-same-fate/ (accessed November 28, 2023).
6. Zott C, Amit R. The fit between product market strategy and business model: implications for firm performance. *Strateg Manag J* 2008;29:1–26. https://doi.org/10.1002/SMJ.642.
7. Sun H, Gilbert SM. Retail price competition with product fit uncertainty and assortment selection. *Prod Oper Manag* 2019;28:1658–73. https://doi.org/10.1111/POMS.13005.
8. Duchak A. In pursuit of new product opportunities: transferring technology from lab to market. *Chem Entrep* 2021:61–101. https://doi.org/10.1002/9783527819867.CH3.
9. Moulos T. Product-market fit: the definitive guide [2023] n.d. https://growthrocks.com/blog/product-market-fit/ (accessed November 28, 2023).
10. Why new products fail & how to prevent it - parlor n.d. https://www.parlor.io/blog/why-new-products-fail-how-to-prevent-it/ (accessed November 28, 2023).
11. Altug MS, Sahin O. Impact of parallel imports on pricing and product launch decisions in pharmaceutical industry. *Prod Oper Manag* 2019;28:258–75. https://doi.org/10.1111/POMS.12908.
12. Cooking 101: making a product launch a reality. *Brand Manag* 101 2012:119–23. https://doi.org/10.1002/9781119207733.CH21.

13. Perrier M, Depeige A. From ideation to product launch. *Innov Engines Entrep Enterp a Turbul World* 2017:91–109. https://doi.org/10.1002/9781119427537.CH5.
14. When corporate innovation goes bad — the 164 biggest product failures of all time n.d. https://www.cbinsights.com/research/corporate-innovation-product-fails/ (accessed November 28, 2023).
15. Zuraikat L. Google glass: a case study. *Perform Improv* 2020;59:14–20. https://doi.org/10.1002/PFI.21919.
16. 17 Terrible product designs that will have you asking, "why?" - CheezCake - Parenting | Relationships | Food | Lifestyle n.d. https://cheezburger.com/8206597/17-terrible-product-designs-that-will-have-you-asking-why (accessed November 28, 2023).
17. 10 Bad web design examples & common errors of website Designers n.d. https://www.mockplus.com/blog/post/bad-web-design (accessed November 28, 2023).
18. 11 Reasons why products fail n.d. https://www.uservoice.com/blog/why-products-fail (accessed November 28, 2023).
19. How bad product design services can cost your company a fortune in fees | Cad Crowd n.d. https://www.cadcrowd.com/blog/how-bad-product-design-services-can-cost-your-company-a-fortune/ (accessed November 28, 2023).
20. Rosenberg NO, Chauvet MC, Kleinman JS. Leading for a corporate culture of design thinking. *Des Think New Prod Dev Essentials from PDMA* 2015:173–86. https://doi.org/10.1002/9781119154273.CH12.
21. Garcia R, Dacko S. Design thinking for sustainability. *Des Think New Prod Dev Essentials from PDMA* 2015:381–400. https://doi.org/10.1002/9781119154273.CH25.
22. Luchs MG. A brief introduction to design thinking. *Des Think New Prod Dev Essentials from PDMA* 2015:1–12. https://doi.org/10.1002/9781119154273.CH1.
23. Szostak BL. Design thinking – design thinking and strategic management of innovation. *Innov Econ Eng Manag Handb 2 Spec Themes* 2021:115–20. https://doi.org/10.1002/9781119832522.CH12.
24. What is design thinking? — updated 2023 | IxDF n.d. https://www.interaction-design.org/literature/topics/design-thinking (accessed November 28, 2023).
25. What is design thinking & why is it important? | HBS Online n.d. https://online.hbs.edu/blog/post/what-is-design-thinking (accessed November 28, 2023).
26. Design thinking 101 n.d. https://www.nngroup.com/articles/design-thinking/ (accessed November 28, 2023).
27. Product Design Process (PDP) n.d. https://engineeringproductdesign.com/knowledge-base/product-design-process/ (accessed November 28, 2023).
28. A comprehensive guide to product design — smashing magazine n.d. https://www.smashingmagazine.com/2018/01/comprehensive-guide-product-design/ (accessed November 28, 2023).
29. A 5 step product design process suitable for all teams n.d. https://www.hotjar.com/product-design/process/ (accessed November 28, 2023).

30. Stages in product design process and development | Boldare - Digital Product Development & Design Company n.d. https://www.boldare.com/blog/digital-product-design-process/ (accessed November 28, 2023).
31. What is product design and the product design process? | by Gloria Lo | UX Planet n.d. https://uxplanet.org/what-is-product-design-and-the-product-design-process-41b41a5bf795 (accessed November 28, 2023).
32. Product design process: steps to designing a product people will love n.d. https://uxstudioteam.com/ux-blog/product-design-process-steps/ (accessed November 28, 2023).

Chapter 7

Visual Design Elements: User Experience (UX) and User Interface (UI) Design

7.1 Visual Design Elements and Principals

The current era has become a fast-paced world where the first impression of a person or a product plays an important role in creating their presence. It is reported that the first impression of anything by a person is formed within 50 milliseconds [1]. Hence, visually, what the person sees first is probably the impression which gets carried on. This is where the role of visual design comes into play. Hence, it becomes important for an entrepreneur to understand the elements and principles of visual design. Visual design involves the creation of an aesthetically pleasing and user-friendly product or service [2, 3]. Essentially, it is a fusion of visual design and user interface (UI) design. Visual design seeks to enhance the aesthetic appeal and usefulness of a product or design by the use of appropriate colour, font, space, and imagery. There is more to visual design than mere aesthetics. Designers strategically position items in order to produce interfaces that maximise conversion and enhance the user experience (UX).

Visual design is essential for both the maker and the consumer of a product [4, 5]. It is crucial for the maker of a product since it influences the purchase decisions of consumers. The ease of use of a product is also determined by its visual design, which is of significance to the end-user. It

DOI: 10.4324/9781003475309-10

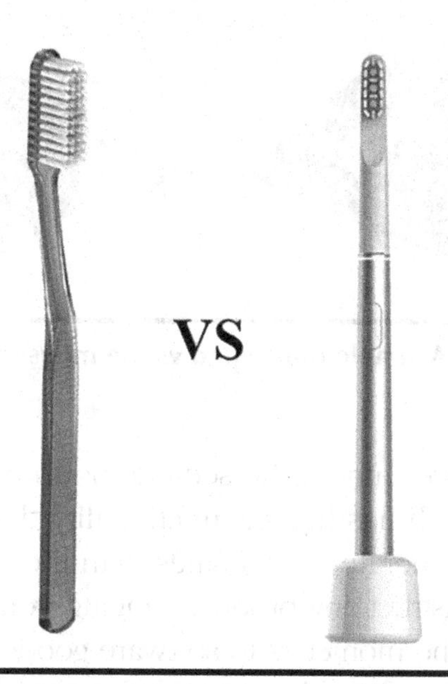

Figure 7.1 Example of a non-aesthetic and an aesthetic product.

not only prevents customers from wasting time and energy learning about the goods, but also mitigates the risk of product damage during usage. As a result, consumers may optimise their experience with the product. The two essential components of visual design are aesthetics and utility. While two goods may perform identically, the majority of individuals will select the more aesthetically pleasing one. Aesthetics encompass a vast array of apparent characteristics, such as form, size, colour, and texture, among others.

Figure 7.1 shows the example of a non-aesthetic and an aesthetic product, i.e., toothbrush. On the left, a simple, and single coloured toothbrush with smooth texture is shown. Whereas, on the right, more aesthetically appealing design is presented. The second toothbrush consists of aluminium handle, a matt textured bristle holder, and an overall sleek design. As per the visual design, customers may tend to buy the second toothbrush just based on its looks. Also, considering a similar functionality (i.e., brushing of teeth), the second product might experience higher sales if the cost is appropriate.

On the other hand, visual communication design is an informal term used to denote visual design. Figuratively, the intention with this notion is to ensure that the product's visual design communicates with the user in order to facilitate use. Consider a novel device that features an altogether new design and functionality. The incorporation of a power button graphic by a visual designer would facilitate the process of activating the gadget for an inexperienced user. From an utility standpoint, this is what visual design comprises. Figure 7.2 shows another example of the functionality. Although

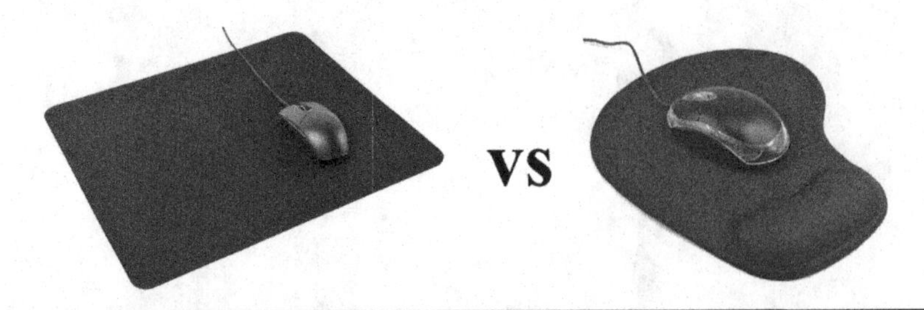

Figure 7.2 Functionality: A simple mousepad vs the mousepad with cushioned wrist support.

the use and aesthetics are similar, the second mousepad consists of a cushioned wrist support which has higher functionality than the left one. Hence, the right product may generate higher sales than the left one.

It is possible to deconstruct any product, ranging from healthcare devices like glucometers and digital thermometers to software goods like hospital websites and medical applications, into its elements that comprise visual design. As visual designers, entrepreneurs rely heavily on these elements in their everyday work; therefore, a fundamental comprehension of the elements is essential. Overall, there are seven design elements as explained by Hashimoto and Clayton [6] in their book titled *Visual Design Fundamentals: A Digital Approach*.

Line: Connecting two points with strokes, lines are the most fundamental component of visual design. By employing them in a repetitive manner, it is possible to generate textures in the form of patterns rather than forms. In the realm of design, diminishing anything below a one-dimensional state is not possible, hence, a line is the basic format that new designers or entrepreneurs could use. Despite their simplicity, lines may exhibit a vast array of characteristics that enable us to communicate a variety of ideas. For instance, lines may exhibit variations in thickness, curvature, straightness, taper, geometric characteristics (resembling those produced by a ruler), or organic qualities (i.e., as if they are drawn by hand). Additionally, a line might be indicated implicitly by the formation of an unseen link between other parts. Figure 7.3 shows an example of a line converted to curved lines. The curved lines could further be used to create company logos. Lines make the logos look minimalistic. In the Figure 7.3, Nestle, Airbnb, and Nike are the examples with lines as their base design for logo creation.

Shape: Self-contained regions, shapes are often made by lines (although they may also be formed by using a different colour, value, or texture). Length and breadth are the two dimensions of a form. Additionally, shapes are fundamental visual design components that frequently comprise the centre of a work of art. Although shapes are

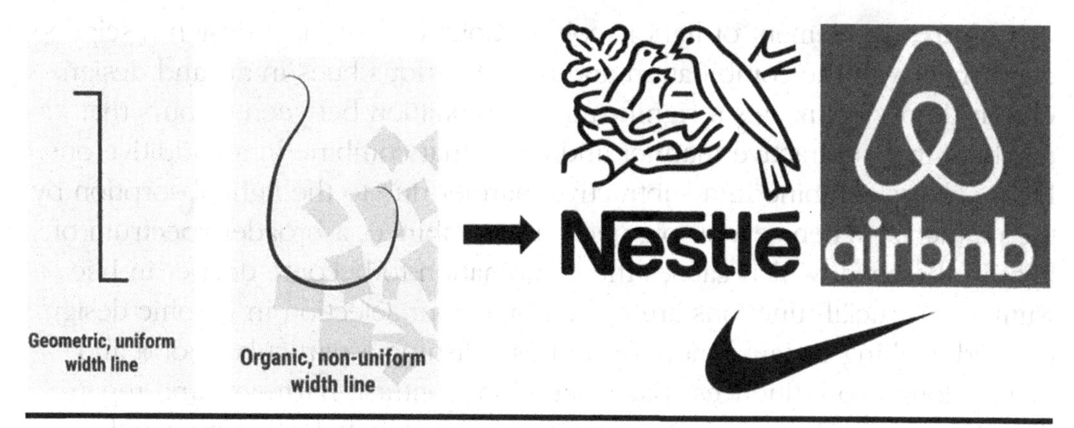

Figure 7.3 Usage of line concept to create logos [1, 7].

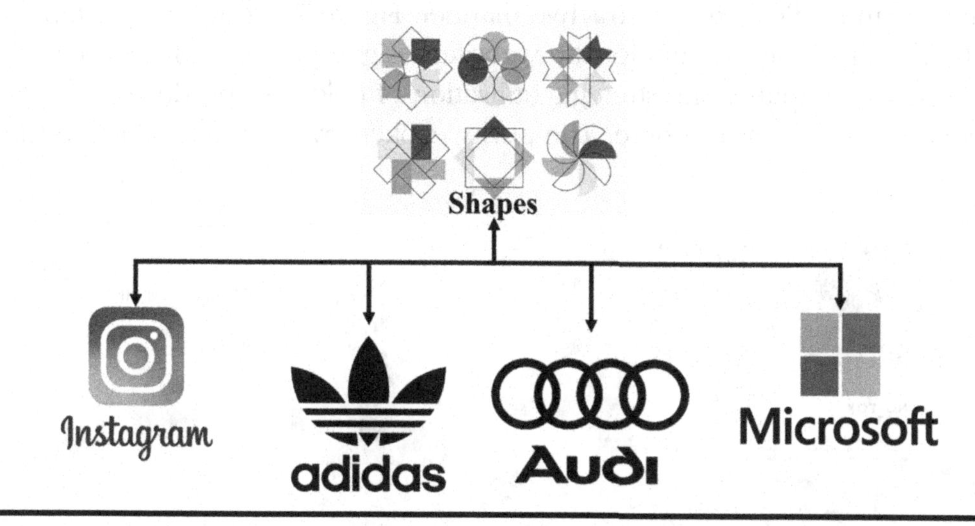

Figure 7.4 Using shapes to generate logo concepts [5].

commonly associated with geometric circles and rectangles, they can take on any conceivable form. Three sorts of forms are utilised in graphic design: mechanical, organic, and abstract. By definition, mechanical forms are geometric in nature, including sharp angles and edges. They are utilised to provide stability and structure to a design. In design, organic forms are those that resemble the natural environment. Texture or rounded edges may be included in order to impart a natural appearance and feel. Abstract forms are irregular configurations that can be included into designs to provide meaning or diversity. Figure 7.4 shows the company logos with geometric shapes as their base. In this example, Instagram is seen to use two squares and circles as its logo. On the other hand, Microsoft and Audi are seen using squares and circles as their abstract design, whereas Adidas is seen using a leaf-shaped design with three rectangular cuts.

Colour: An element of light is colour. Colour theory is a design discipline concerned with the combination and use of various hues in art and design. Colour theory delineates a significant differentiation between colours that combine in a subtractive manner and those that combine in an additive one. Paint colours combine in a subtractive manner due to the light absorption by the pigments. When several pigments are combined, a broader spectrum of light is absorbed, which causes the combination to become deeper in hue. Numerous crucial functions are served by colour selection in graphic design. Beyond evoking certain emotions and establishing a particular mood and tone, colour also influences space, variation, contrast, harmony, and repetition. As a result of these numerous functions, colour selection may make or break a design. A black hue is generally produced by combining cyan, magenta, and yellow in a subtractive manner. Figure 7.5 shows an example of the use of simple colours for the website designed by Neil [8]. Neil's website design exemplifies the strategic utilisation of colour to produce a distinct aesthetic and atmosphere. The purity evoked by the white space in the

Figure 7.5 Example of usage of colours to design websites [8].

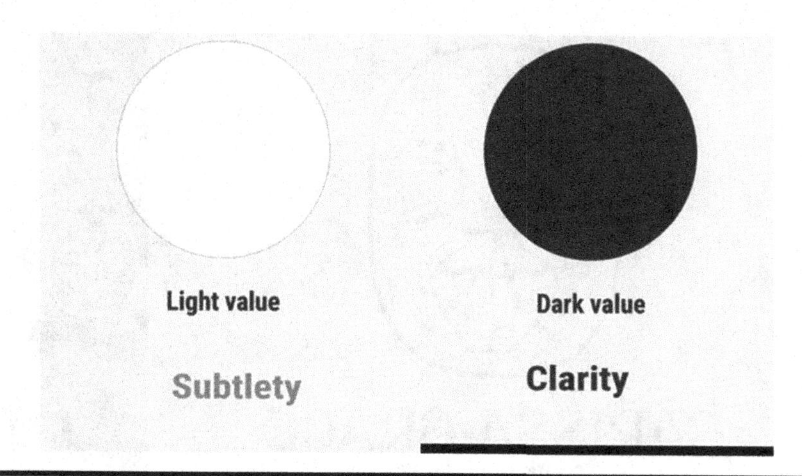

Figure 7.6 Understanding the value of light and dark through visual design [1, 7].

design is further complemented by the brand's natural approach to beauty and wellbeing, which is reflected in the gentle beige and green hues.

Value: The value of a colour denotes its degree of brightness or darkness. The significance of this component in visual design is in the contrast produced by the juxtaposition of bright and dark values. A design characterised by a significant contrast of values—specifically, the utilisation of bright and dark values—engenders a perception of clarity, whereas a design employing comparable values imparts an impression of subtlety. Additionally, volume may be simulated in two dimensions using values; for instance, brighter values might be employed to indicate where light strikes the object, while darker values correspond to shadows. Figure 7.6 shows the light and dark values where the text feels faded and clear respectively.

Texture: There are two distinct categories of textures that a designer may employ: tactile textures, which are perceptible through the sense of touch, and inferred textures, which are only seen without the ability to feel. Tactile textures consist of surfaces that can be touched, including harsh sandpaper and soft hair. There are several motives and methods for incorporating texture into a design. Texture has the ability to impart a sense of organic roughness and authenticity to a design, even if it is entirely digital in nature. Texture visually provides depth and may enrich and diversify a design. Nevertheless, texture should be employed sparingly so as not to dominate the design and become overpowering. Figure 7.7 shows the design by Jesse Bowser [9]. It is an exquisite illustration of the way in which visual texture can enhance a design. This example's texture complements the natural and tactile nature of the brand.

Figure 7.7 Texture used in design [9].

Figure 7.8 Logo design using space as a concept [10].

Space: Space refers to the region encompassing or separating components inside a design. In design, "white space" is a common term for space; however, it does not need to be white. Additionally, space can refer to negative space or the area between forms. Sava Stoic's [10] ingenious logo for the Homeleon brand combines the forms of a chameleon and a house through the utilisation of negative space (Figure 7.8). Space is an essential component of any design, unless the intention is to achieve an absolute state of disorder. Architectural design components are afforded sufficient breathing room by means of space. This also applies to typography: it is challenging to read letters that are densely clustered. As well as serving to partition components, space may also contribute to the equilibrium of a design by accentuating particular parts.

Volume: When referring to three-dimensional things in a design, volume or shape is used. Volumetric illusion is the sole sort of volume utilised by visual designers, given that visual design is a two-dimensional style of design. That is to say, visual designers have the ability to provide the appearance of volume by utilising 3D images and forms. Figure 7.9 shows an example of a webpage utilising the concept of volumes to create a perception for 3D.

Figure 7.9 Designing abstracts using the volume concept [11].

Line, form, negative/white space, volume, value, colour, and texture are the components of visual design that provide the aesthetic foundation of a product. Conversely, the principles of design delineate the optimal combinations and interactions of these components. A number of the following principles are complementary and closely linked. The seven principles of visual design are explained below [12].

Scale: The idea of scale defines signalling importance and hierarchy in a composition through the use of relative size. Put simply, when this concept is used appropriately, the parts that hold the utmost significance in a design are rendered more substantial than those that do not. The rationale for this theory is straightforward: the greater the size, the greater the probability of noticing. In general, a visually appealing design employs no more than three distinct sizes. Incorporating a number of items with varying sizes into the layout can not only add visual diversity, but also construct a visual hierarchy. By effectively applying the idea of scale and emphasising the appropriate aspects, users may be capable of comprehending and utilising the image without difficulty.

Visual Hierarchy: The concept of visual hierarchy pertains to the method of directing the reader's attention across the page, prioritising distinct design components according to their relative significance. Diverse signals, including differences in magnitude, value, colour, spacing, and position, can be utilised to establish visual hierarchy. Visual hierarchy dictates the

Figure 7.10 An example of balance [12].

manner in which the experience is delivered. When navigating a webpage and finding it difficult to discern where to look, it is most probable that the layout lacks a distinct visual hierarchy. Utilise two to three typeface sizes to convey to visitors which elements of the page's mini-information architecture are more vital or at the highest level, so establishing a distinct visual hierarchy. Conversely, contemplate employing subdued hues for less significant things and vibrant tones for critical ones.

Balance: The concept of balance pertains to the harmonious organisation or ratio of components in a design. When a quantity of visual input is dispersed evenly (although not necessarily symmetrically) on both sides of an imaginary axis passing through the centre of the screen, balance is achieved. Although frequently vertical, this axis can also be horizontal. Comparable in nature to the process of weight balance, a design with a single large and one small feature on opposite sides of the axis would impart a sense of instability. In achieving balance, it is not just the quantity of pieces that matters, but also the area occupied by the design element.

Consider the example in Figure 7.10. This page is asymmetrically balanced, which is consistent with the Nike brand by conveying a sense of motion and vitality. Drawing a vertical axis in the middle of this graphic would reveal that the number of items on both sides of the axis is approximately equivalent. Nevertheless, they differ in that they are not similar while being situated in the exact same areas.

Figure 7.11 Gestalt's principle used in the NBC logo [12].

Contrast: The notion of contrast entails the deliberate arrangement of visually contrasted pieces so as to communicate their distinction (e.g., belong in different categories, have different functions, behave differently). In other words, contrast emphasises the distinction between two sets of things or two objects by providing the eye with a visible difference (e.g., in colour or size) between them. Frequently, the idea of contrast is implemented using colour.

Gestalt Principles: Consciously organising the components into a systemic whole rather than viewing them as a succession of isolated aspects is how humans simplify and organise complicated visuals comprised of several elements, according to Gestalt principles. Put simply, Gestalt principles encompass our inclination to view the entirety rather than its component parts. Similarity, continuity, closure, closeness, common region, figure/ground, symmetry, and order are among the numerous Gestalt principles. For example, while the white space does not contain a peacock, the human brain perceives the presence of one in the NBC logo (Figure 7.11).

7.2 User Experience (UX) and User Interface (UI) Design

User experience (UX) design is the method by which design teams develop products that offer users relevant and significant experiences. UX design encompasses the vision for the complete product acquisition and integration process, including branding, design, usability, and functionality. Experience design includes not just the development of user-friendly software, but also the creation of supplementary experiences associated with the product, such as packaging, marketing campaigns, and post-purchase assistance. Primarily, UX design prioritises the provision of solutions that effectively tackle challenges and requirements. In the end, a product that serves no purpose will be abandoned [13].

Designers have little influence on an individual's perceptions and reactions—the initial aspect of the definition—in UX. For instance, they lack the ability to regulate an individual's emotions, finger movements, or eye movements when utilising a product. However, designers have the ability to influence the appearance and behaviour of the product, system, or service—the second component of the definition. UX design may be conceptualised in its most basic form as both a verb and a noun. The elements that influence the user experience (noun)—perceptions and reactions to a system or service—are designed by a UX designer.

For instance, certain characteristics of a physical object, such as a computer mouse, can be modified to affect the user's enjoyment of its appearance, feel, and hold: The manner in which it suits their hand. Does it fit snugly? Is it cumbersome and too large? The quantity. Does it influence their capacity to manipulate it as desired? Its user-friendliness. Is it automatic to use, or do they need to exert effort in order to accomplish a task? When an individual utilises a digital product, such as a computer programme, there are several facets that fall under control: he degree to which they can traverse the system intuitively. The indicators that direct them toward their objective. The timely and adequate visibility of the critical elements associated with a given task.

UX design is a heterogeneous subject due to the fact that it spans the complete user journey. UX designers have many backgrounds, including programming, psychology, graphic design, and interaction design. In addition to working with a broader scope regarding accessibility, designing for human users necessitates considering the physical constraints of many potential users, such as the inability to see small writing.

The duties of a UX designer are many, but frequently encompass user research, persona development, wireframe and interactive prototype creation, and design testing. The responsibilities associated with these duties may differ substantially among organisations. However, they consistently need designers to speak for users and ensure that their requirements remain central to all design and development endeavours. This is another reason why the majority of UX designers employ a user-centred work method, wherein they continue to direct their most informed efforts until they have effectively resolved all pertinent concerns and user requirements. Figure 7.12 shows the cycle for user-centric design.

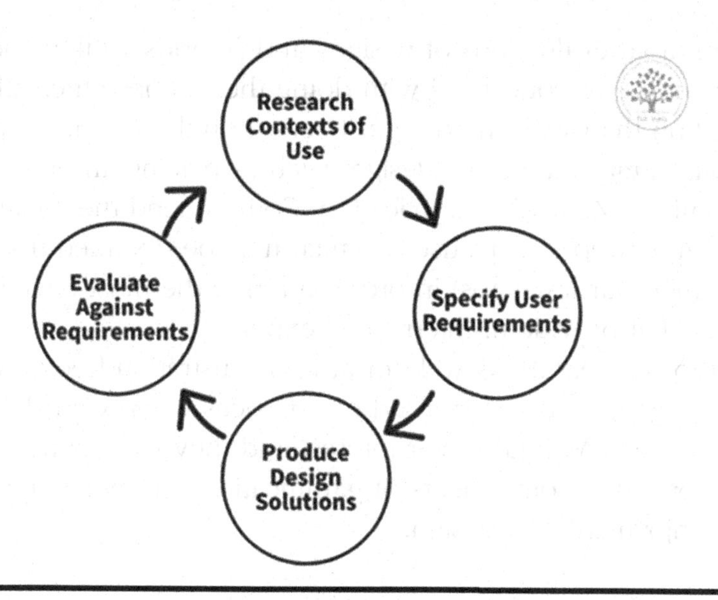

Figure 7.12 Process cycle of a user-centric design [13].

Figure 7.13 Three formats of user interface designing [14].

User interface (UI) design is the method by which programmers and device manufacturers construct interfaces with an emphasis on aesthetics and design. The objective of designers is to provide user interfaces that are intuitive and enjoyable. UI design encompasses several sorts of user interfaces, including voice-activated interfaces. User interfaces serve as the means by which individuals engage with designed elements [14]. There are three forms available (Figure 7.13).

Graphical user interfaces (GUIs)—on digital control panels, users interact with visual representations. The desktop of a computer is a GUI. VUIs means voice-controlled user interfaces The voices of users are utilised to communicate with them. The majority of smart assistants are VUIs, including Alexa on Amazon devices and Siri on iPhones. Gesture-based interfaces—in virtual reality (VR) games, for instance, users interact with three-dimensional design areas using their body movements.

Users form rapid evaluations of designs and prioritise utility and aesthetic appeal. They are more concerned with doing their duties efficiently and with minimal exertion than with the design. Users should concentrate on accomplishing activities rather than the design, such as placing an order for pizza using the Domino's Zero Click application. Comprehend the contexts and task flows of an entrepreneur's users (which may be obtained from sources such as customer journey maps) in order to refine the most effective and user interfaces that provide uninterrupted experiences.

UIs ought to be pleasant as well (or at least frustration-free). Users may have more engaging and personalised experiences when your design anticipates their demands. Maintain their interest and they will continue to patronise the entrepreneur. Components of gamification can, when applicable, enhance the enjoyment of a design.

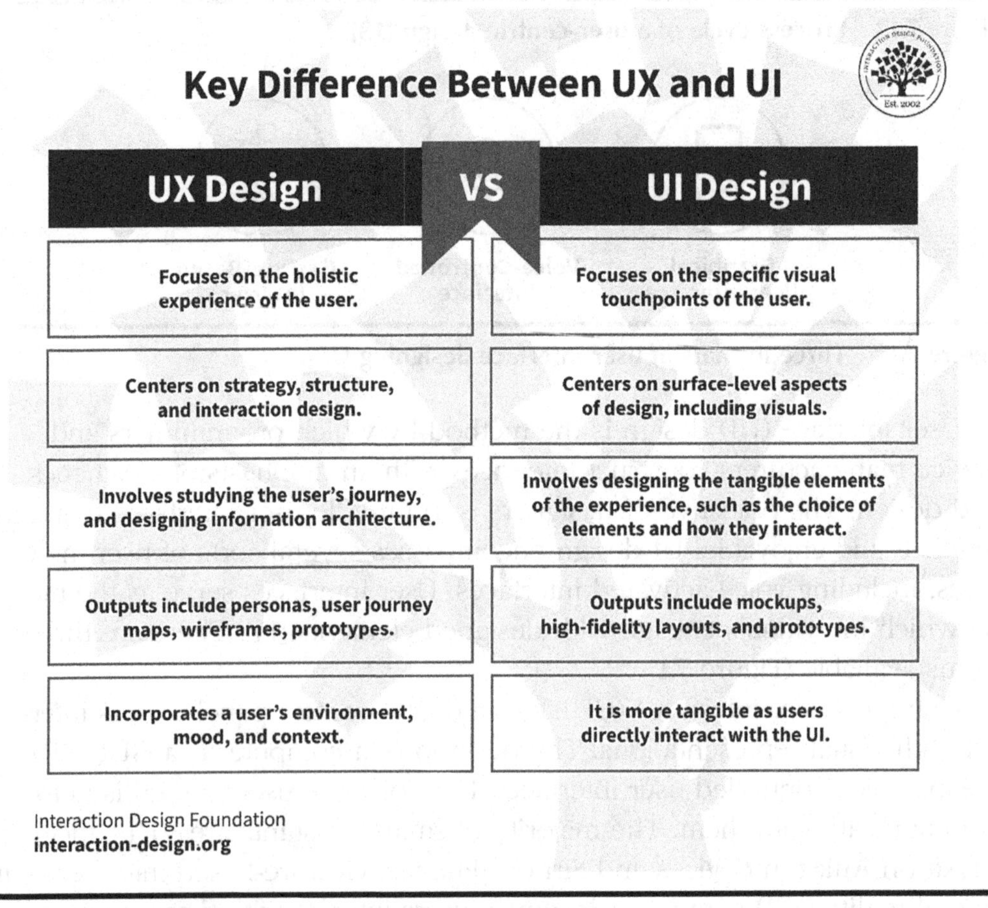

Figure 7.14 Differences between UX and UI [15].

Understanding UX and UI is fundamental to the user design methodology. UX guarantees that the end-user receives value in product design, whereas UI emphasises efficiency, effectiveness, and customer pleasure. Consider, for example, mobile applications. A cohesive integration of UX and UI guarantees seamless user navigation and an aesthetically pleasing presentation. A designer that comprehends UX may create intuitive user routes to circumvent poor UX. A good UX in an e-commerce application, for instance, facilitates navigating among product categories. Conversely, UI expertise guarantees the utilisation of impactful visual components that are most suitable for any undertaking in order to elevate the overall user experience. User experience design aims to detect and resolve user issues whereas, UI is concerned with developing aesthetically pleasing, engaging, and user-friendly interfaces [15]. Figure 7.14 represents the difference between UX and UI in detail.

References

1. The key elements & principles of visual design | IxDF n.d. https://www.interaction-design.org/literature/article/the-building-blocks-of-visual-design (accessed December 3, 2023).
2. Jongmans E, Jeannot F, Liang L, Dampérat M. Impact of website visual design on user experience and website evaluation: the sequential mediating roles of usability and pleasure. *J Mark Manag* 2022;38:2078–113. https://doi.org/10.1080/0267257X.2022.2085315.
3. What is visual design? | CorelDRAW n.d. https://www.coreldraw.com/en/tips/visual-design/ (accessed December 3, 2023).
4. Kostelnick C. Humanizing visual design: the rhetoric of human forms in practical communication. *Humaniz Vis Des Rhetor Hum Forms Pract Commun* 2019:1–280. https://doi.org/10.4324/9781315114620/HUMANIZING-VISUAL-DESIGN-CHARLES-KOSTELNICK.
5. The visual design elements and principles that make good design n.d. https://www.flux-academy.com/blog/the-visual-design-elements-and-principles-that-make-good-design (accessed December 3, 2023).
6. Hashimoto A.Clayton M. Visual design fundamentals: a digital approach, (3rd. ed.) 2009. Charles River Media, Inc., USA.
7. What is visual design? — updated 2023 | IxDF n.d. https://www.interaction-design.org/literature/topics/visual-design (accessed December 3, 2023).
8. Botanika | beauty store by Neil on Dribbble n.d. https://dribbble.com/shots/11632805-Botanika-Beauty-Store (accessed December 3, 2023).

9. Carissa grace floral design by Jesse Bowser on Dribbble n.d. https://dribbble.com/shots/10591878-Carissa-Grace-Floral-Design (accessed December 3, 2023).
10. Homeleon - Ultimate Logofolio entry by Sava Stoic on Dribbble n.d. https://dribbble.com/shots/22086289-Homeleon-Ultimate-Logofolio-Entry (accessed December 3, 2023).
11. Timescope by Peter Tarka on Dribbble n.d. https://dribbble.com/shots/13991690-Timescope (accessed December 3, 2023).
12. 5 Principles of visual design in UX n.d. https://www.nngroup.com/articles/principles-visual-design/ (accessed December 3, 2023).
13. What is user experience (UX) design? — updated 2023 | IxDF n.d. https://www.interaction-design.org/literature/topics/ux-design (accessed December 3, 2023).
14. What is user interface (UI) design? — updated 2023 | IxDF n.d. https://www.interaction-design.org/literature/topics/ui-design (accessed December 3, 2023).
15. UX vs UI: What's the difference? | IxDF n.d. https://www.interaction-design.org/literature/article/ux-vs-ui-what-s-the-difference (accessed December 3, 2023).

Chapter 8

Quality Engineering and Iterative Design Optimisation for Healthcare Products

8.1 Quality Engineering

The primary objective of quality engineering is to ensure that the design, development, and production of goods and services conform to or exceed the expectations and demands of consumers [1]. It encompasses all tasks associated with the evaluation of the design and development of a product. Additionally, quality engineers ensure that the items are manufactured in accordance with the requirements. A collection of approaches, procedures, and concepts known as quality engineering assists a company in enhancing its business operations in order to attain a specified standard of quality for its offerings. Figure 8.1 shows the different aspects of quality, namely performance, reliability, durability, features, conformance, serviceability, aesthetics, and perceived quality.

A quality engineer is concerned with waste reduction in addition to the quality of the product and the manufacturing process. Process quality is usually designed and monitored by quality engineers [2]. They operate across diverse sectors and fulfil a critical function by rectifying or resolving flaws. Hence, both quality engineering and assurance are important. Quality engineering and quality assurance have a few differences. Quality assurance comprises all procedures that verify that a product is manufactured correctly. For this to occur, the manufacturer needs a quality management system [3]. Conversely, quality engineering

DOI: 10.4324/9781003475309-11

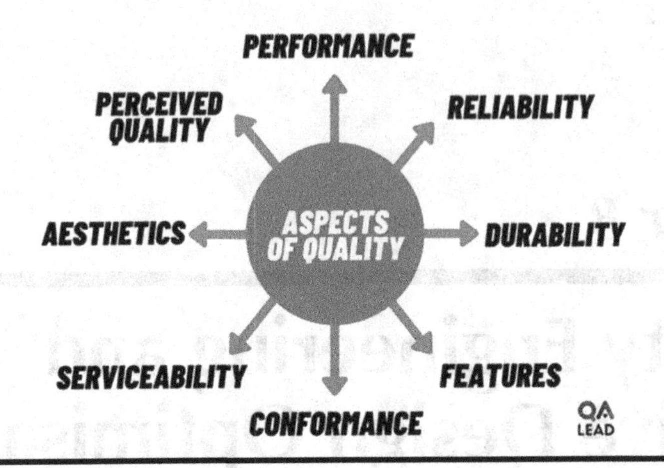

Figure 8.1 Different aspects of quality [2].

establishes the parameters of such a system. Furthermore, it enhances and maintains it. Quality engineers "engineer" the manufacturing process, whereas quality assurance is concerned with preserving quality. Quality engineers, to put it another way, design the system. Certain quality engineers are tasked with both the design and monitoring of the system. In addition to this, feedback of a customer is also important for quality control [4].

By ensuring that the voice of the customer is incorporated into new or upgraded goods and services and facilitating improved communication between the engineering and manufacturing teams, a company may achieve an effective application of quality engineering processes and tools [5]. Tools for quality engineering can increase the quality of digital products while decreasing expenses via more streamlined methods. Effective implementation is crucial for developing superior digital goods that not only satisfy but also exceed the expectations of consumers, which can have a significant impact on financial performance [6].

To understand the trends of quality engineering in healthcare, the following example has been stated. The majority of users employ online banking or retail services such as Amazon, Walmart, or Alexa (IoT devices) to place online orders that are delivered the same day to their doorstep or to their kerbside. Additionally, physicians and patients are consumers who need information readily available via their mobile devices and speedier internet connections [7].

Furthermore, the healthcare sector has been profoundly affected by the unusual worldwide pandemic of COVID-19, which has affected all industries in unexpected ways. Several gaps remain in healthcare systems' ability to deliver a digital health experience to patients and hospital staff at any time and in any location, despite the efforts of many frontline workers and hospitals to save lives. Despite the conclusion of the pandemic, the healthcare industry will continue to operate without adequately addressing the expanding digital demands. Thus, in order to make healthcare more accessible, user-friendly, and manageable, hospitals and health systems require a strategy to re-engineer their digital transformation. Furthermore, there is a trend toward "value-based health," which entails improved outcomes, quality of care for both patients and service providers, and cost reduction. Ultimately, patients will feel far closer to their physicians, and technological advancements are propelling this trend.

Telehealth Services: The COVID pandemic has taught the requirement for telehealth monitoring. Although telehealth services are not a novel technology, their implementation has been somewhat sluggish. In light of the ongoing social distance issue, the ability of healthcare professionals to deliver virtual care and enable physicians and nurses to communicate with patients via email, phone, and video-chatting has become an absolute necessity. The Federal Communications Commission (FCC) stated on April 2, 2020, that it will award qualifying healthcare providers with $200 million in Congress-appropriated funds under the COVID-19 Telehealth Program. To enable the expansion of telehealth, the IT sector will need to discover, create, and implement a variety of secure patient–physician communication systems. Telehealth services integrated with chatbot functions provided by artificial intelligence (AI) will enhance the quality of treatment for all patients, functioning as retail-like healthcare services that continue to operate beyond the conclusion of social distance. Reengineering software testing is more important than for any other application category in the telehealth industry since lives depend on apps and a single error can result in permanent impairment or death for a patient [7].

Remote Health Monitoring: Healthcare services in which wearables and health applications assess patients' data in real time and transmit crucial, real-time information would be significantly impacted. Patients have the ability to oversee various aspects of their health, including sleep patterns, heart rate, and chronic conditions such as diabetes or blood sugar, by utilising remote health monitoring with user-friendly devices that are equipped with Internet of Things (IoT) technology and Bluetooth capabilities. Physicians are also well positioned to comprehend patients' conditions and provide prompt treatments [7].

Pharmaceutical Domain: Integral to the pharmaceutical industry has always been the quality engineering procedures of computerised systems, which include validation and verification in addition to quality assurance and control [8]. It has effectively safeguarded customers from any potential harm resulting from substandard manufacturing procedures within the sector. Additionally, this procedure motivates quality assurance testing engineers in pharmaceutical organisations to strive for enhanced operational efficiency and decreased expenses. Quality standards maintenance in the pharmaceutical sector has consistently been enhanced by technology advancements. Research from 2021 confirms this and emphasises how AI, Big Data, Robotics, IoT, and Cloud Computing, among others, are upending the established technological landscape in the pharmaceutical sector while also contributing to its overall efficiency [9]. Nevertheless, in order to operate efficiently and support the pharmaceutical industry, the majority of these technologies must undergo software quality assurance.

Audits and regulations are stringent in the pharmaceutical, healthcare, and medical device sectors, which makes compliance a significant liability for all activities. The quality of software utilised in the production, quality control, and documentation of pharmaceuticals is assessed in relation to standards established by governmental organisations, such as the US Food and Drug Administration (FDA). The primary objective of the FDA is to avert the dissemination of potentially dangerous drugs that have been tainted or mislabelled due to software defects. In addition, data integrity is a significant challenge for business, quality control, and production systems.

Medical Devices: The healthcare sector is now experiencing a paradigm shift due to the development and implementation of novel medical devices and therapies [10]. Medical gadgets have emerged as critical components in facilitating this paradigm shift, which would provide enhanced connection for the seamless execution of data analytics and remote patient monitoring. In order to adapt to the ever-evolving market dynamics, medical devices must remain at the forefront of innovation and make use of cloud computing, mobility, analytics, and wireless technology. There is an urgent requirement for medical equipment to enhance the monitoring of patients' healthcare parameters, facilitate precise diagnosis of illnesses, and optimise expenses, among other functions. In order to guarantee optimal performance, software testing should be the standard throughout for medical equipment. Additionally, organisations should personalise the appearance of these devices to ensure quality compliance and user-friendliness [11].

The significance of medical technologies in relation to human life cannot be emphasised enough. In order to guarantee the efficacy, precision, and safety of healthcare services, medical equipment has to undergo exhaustive testing. It is imperative that it undergoes a thorough verification and validation procedure in order to ensure the provision of dependable and high-quality outputs. In spite of this, medical device producers should persist on utilising quality engineering due to the industry's competitive nature. This is because the Voice of the Customer may be effectively included into the product design through the usage of quality engineering services. Effective implementation of quality engineering methods and technologies may have a substantial impact on the cost and quality of medical devices.

Quality engineering surpasses quality assurance in that it guarantees that the design and production of medical equipment not only satisfies but also exceeds the expectations of clients. An approach is used where quality is integrated into the design and development phases, minimising costs by identifying bugs in the early stages of development. Utilising a cross-functional approach, quality engineering for medical devices encompasses an extensive array of approaches and tools. Given that medical devices are required to adhere to the most stringent regulatory standards and are anticipated to provide outstanding performance, safety, and efficacy for end-users, testing for such devices should be guided by quality engineering principles. The procedure needs to start with the design stage and go through the manufacturing stage. Aside from others, the quality assurance plan for medical devices should assure adherence to rules such as those of the FDA and other organisations.

8.2 Iterative Design Optimisation

The iterative design process is a straightforward principle. A prototype is created subsequent to the identification of a user need and the generation of ideas to fulfil that need via user research. The prototype is then put to the test to see whether it optimally satisfies the requirement. Then, based on the insights gained from testing, modifications are made to the design. Following that, a new prototype is created and the iterative process is repeated until the user is content that the final product is optimal for commercialisation. This iterative procedure is frequently referred to as "spiral prototyping" or "rapid prototyping" [12]. Figure 8.2 represents the overall steps of a product's or service's design iteration process. It requires an initial planning and understanding of requirements to start the design procedure. After this, evaluation of the product, followed by further planning, analysis, and implementation could be conducted. Post this, testing and deployment of the product could be performed. Based on this, feedback on the final product through the voice of customer is required for further iterations.

It is possible to implement iterative design at any stage of the design process, even after the product has been released to the market and

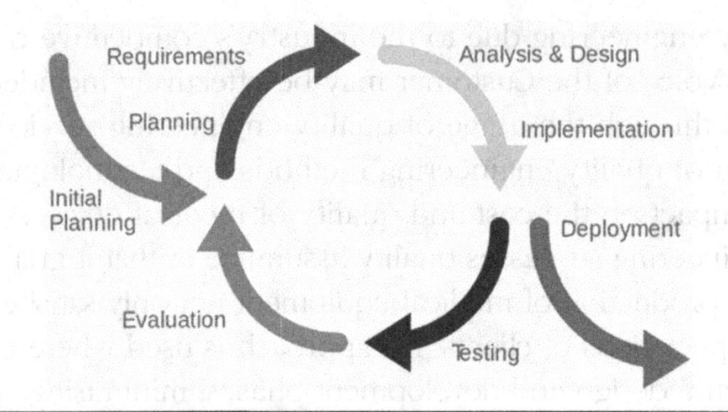

Figure 8.2 Process for iterative design optimisation [12].

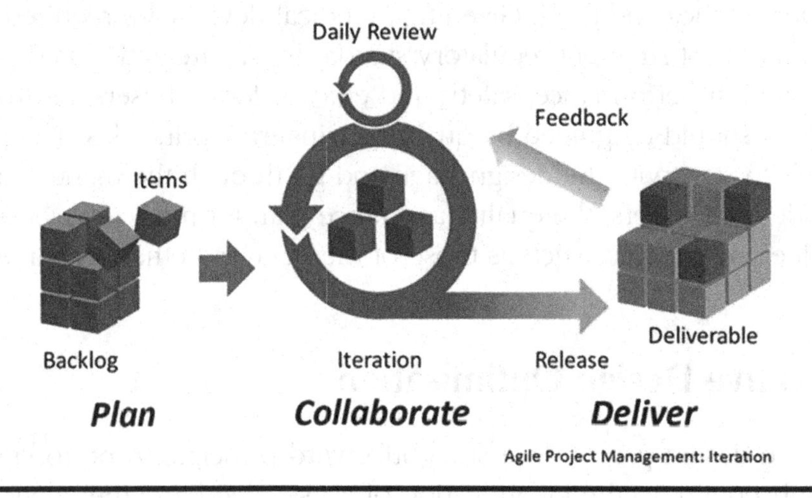

Figure 8.3 Project management technique as an iterative design optimisation process [12].

enhancements are desired. But it is important to note that using iterative design sooner in a product's lifecycle will result in a more cost-effective strategy [13]. Figure 8.3 represents the project management technique which could also be applied to iterative optimisations in a healthcare product of service. An entrepreneur could analyse their backlogs and consider a daily review of their products. Following the feedback and daily review, necessary iterations could be carried out.

To implement the iterative design process, the following principles could be followed [14].

Acquiring Feedback: Throughout the product development process, input is gathered from stakeholders or consumers. This input may have qualitative details on aspects that are effective, in addition to challenges or deficiencies.

Understanding Results: The results obtained from the collection of feedback are meticulously examined. By analysing user interactions with the product, designers are able to spot any issues that may arise.

Modifying the Design: In light of the outcomes analysed, designers implement alterations to the design of the product. Potential actions to consider include interface modifications, feature additions, or design element adjustments.

Reiterating the Process: The iterative process of collecting input, doing analysis, and making adjustments persists over several iterations. Every subsequent iteration strives to enhance the product's refinement.

A significant advantage of iterative design is that it inherently requires extensive user testing. The accumulation of this testing constitutes a vast repository of user input that can be utilised to enhance the design, usability, and ultimately the consumer experience of your website. User feedback and testing may assist you in determining which components of your website are functioning properly and which are not. It can also assist in identifying the underlying source of issues that you may be aware of but unsure of their origin [15]. For instance, in the event that an ecommerce website has a notable increase in abandoned visits at a specific stage of the user journey, the proprietors of the site probably possess knowledge about the location of the issue, without insight into its cause. In this regard, user testing may be quite beneficial.

Prototyping, testing, and refining are excellent methods for identifying possible issues early in the design process, as opposed to at the very end. This may prevent the need for an expensive and time-consuming overhaul in the future. Furthermore, while identifying issues, the iterative design approach can also provide a distinct trajectory for enhancement. The overall experience will become more useful as it undergoes more testing and refinement. Iterative design enhances usability in several aspects, such as the time required to perform task scenarios, the overall user satisfaction rate, and the quantity of usability issues [16].

Considering the extensive testing and ongoing refinement involved, iterative design may appear to be an expensive and time-consuming process. However, it is considerably more efficient in two aspects. To begin with, iterative design reduces the amount of time spent creating papers that delineate and depict the design in comparison to the conventional waterfall methodology. By designing incrementally, you eventually invest additional time in the process of developing and creating the product. The second benefit of iterative design is that it facilitates more efficient issue resolution and maintenance. The need for a comprehensive redesign of their website could be prevented by continuously making adjustments and enhancements. Conversely, iterative design facilitates the incorporation of necessary modifications as they arise [17].

Consider the example of Wikipedia, which is comprised of user-generated material. At any moment, everyone is welcome to contribute to the improvement of the content. A reviewer (or editor) can readily identify the improvement in question and determine if it merits the change or detracts from anything else. According to this notion, Wikipedia's material ought to develop over time in order to establish itself as the preeminent online encyclopaedia. It is important to note that the design iteration strategy is difficult. Given the vastness of society, an individual's "improvement" might very well be an individual's "detriment." Compared to Wikipedia, you will likely have greater influence over the decision-making process if you employ an iterative approach in your own design process.

Another example, related to medical device design is an inhaler used by asthmatic patients. The device is K-haler® developed by Mundipharma [18]. As reported by the manufacturer, it enables patients to inhale gently, circumventing the coordination challenges commonly associated with existing inhalers and the required inhaling force. This facilitates the coordination of inhalation and respiration by the patient, which eventually enhances medication administration. In addition to this, they provide an interactive web training platform, comprising gamified training sequences and step-by-step video tutorials, to enable patients to observe and operate the new K-haler® device, in addition to the device, pack, and patient information leaflets. Figure 8.4

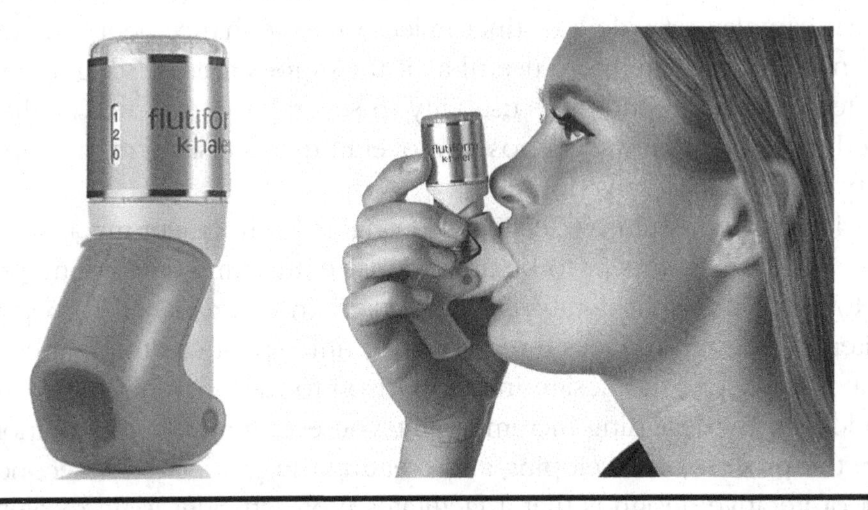

Figure 8.4 K-haler® developed through the iterative design optimisation technique [18].

shows the final product developed through the iterative design optimisation process. The designer considered feedback and noted down the problems. Keeping the key difficulties in mind, major changes were applied to the device's design, leading to the final product.

References

1. What is quality engineering? Definition and examples n.d. https://marketbusinessnews.com/financial-glossary/quality-engineering/ (accessed December 10, 2023).
2. What is quality engineering? - The QA lead n.d. https://theqalead.com/quality-engineering-planning-strategy/what-is-quality-engineering/ (accessed December 10, 2023).
3. Zonnenshain A, Kenett RS. Quality 4.0—the challenging future of quality engineering. *Qual Eng* 2020;32:614–26. https://doi.org/10.1080/08982112.2019.1706744.
4. Quality engineering services for health insurance n.d. https://www.coforge.com/resource-library/data-sheets/quality-engineering-services-for-health-insurance (accessed December 10, 2023).
5. Bergman B, Hellström A, Lifvergren S, Gustavsson SM. An emerging science of improvement in health care. *Qual Eng* 2015;27:17–34. https://doi.org/10.1080/08982112.2015.968042.
6. What is quality engineering and why is it important? - Digivante n.d. https://www.digivante.com/blog/quality-engineering-important/ (accessed December 10, 2023).
7. Quality engineering trends in healthcare n.d. https://rcgglobalservices.com/blog/quality-engineering-trends-in-healthcare (accessed December 10, 2023).
8. Bisgaard S, Does RJMM. Quality quandaries: health care quality reducing the length of stay at a hospital. *Qual Eng* 2009;21:117–31. https://doi.org/10.1080/08982110802529612/ASSET//CMS/ASSET/05211F8F-6016-4852-8E6C-439AA14D9329/08982110802529612.FP.PNG.
9. Why quality engineering has become a top priority for Pharma companies n.d. https://www.qualitestgroup.com/insights/blog/why-quality-engineering-has-become-a-top-priority-for-pharma-companies/ (accessed December 10, 2023).
10. Quality engineering in healthcare: 3 keys to happy customers | tapQA n.d. https://www.tapqa.com/quality-engineering-healthcare-3-keys-happy-customers/ (accessed December 10, 2023).
11. Quality engineering for medical devices | EuroSTAR huddle n.d. https://huddle.eurostarsoftwaretesting.com/quality-engineering-for-medical-devices/ (accessed December 10, 2023).
12. Design iteration brings powerful results. so, do it again designer! | IxDF n.d. https://www.interaction-design.org/literature/article/design-iteration-brings-powerful-results-so-do-it-again-designer (accessed December 10, 2023).

13. Ni M, Borsci S, Walne S, Mclister AP, Buckle P, Barlow JG, et al. The Lean and Agile Multi-dimensional Process (LAMP) – a new framework for rapid and iterative evidence generation to support health-care technology design and development. *Expert Rev Med Devices* 2020;17:277–88. https://doi.org/10.1080/17434440.2020.1743174.
14. What is iterative design? (and why you should use it) | Enginess industry insights: stay ahead with the latest trends and strategies n.d. https://www.enginess.io/insights/what-is-iterative-design (accessed December 10, 2023).
15. Iterative design: how to optimize the product design process | by Vladimir Pavlov | Oct, 2023 | Medium n.d. https://medium.com/@vpavlov.me/iterative-design-how-to-optimize-the-product-design-process-594f955fa398 (accessed December 10, 2023).
16. What is iterative design and how does it work? - Monarch innovation private limited n.d. https://www.monarch-innovation.com/what-is-iterative-design-and-how-does-it-work (accessed December 10, 2023).
17. Gupta SP, Pidgeon A. An analytical approach to identifying potential use-related issues concerning a medical device under development. *J Med Eng Technol* 2016;40:61–71. https://doi.org/10.3109/03091902.2015.1132785.
18. Mundipharma | Healthcare device, UX UI, brand & packaging design | Recipe Design n.d. https://recipe-design.com/work/project/flutiform-mundipharma (accessed December 10, 2023).

Chapter 9

Design for Healthcare Product Manufacturing

9.1 Design for Manufacturing and Industrial Design

The manufacturing industry has seen a radical transformation during the past several decades. As a result of technological advancements and supply chain improvements, businesses may now procure components from all over the world. This has resulted in increased competition, which has pushed businesses to produce new goods in order to expand their consumer bases and reduce profit margins. Consequently, organisations are embracing the Design for Manufacturing and Assembly (DFMA, DFM/A, or DFM/DFA) methodologies, which empower the rapid and cost-effective development of superior goods [1].

Combining the DFM and DFA techniques (DFMA) [2] is a fundamental approach utilised in the creation of medical device products. It facilitates the incorporation of manufacturing process concerns into feature design, including aspects such as material, geometry, tooling, techniques, injection, etc. [3]. This amalgamation facilitates streamlined production and straightforward assembly of a product blueprint [4]. By transforming the sequential process of developing and manufacturing, DFMA fosters a greater sense of collaboration. Figure 9.1 represents the flow chart process differentiating DFM and DFA.

The task of overseeing production involved the design engineer preparing product drawings in isolation before transferring them to the individual responsible for manufacturing. The determination of the production and assembly methods would occur at this phase. DFMA, on the other hand,

DOI: 10.4324/9781003475309-12

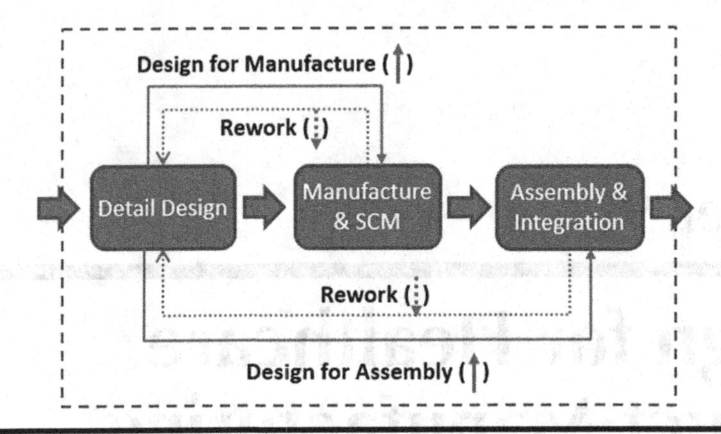

Figure 9.1 Flow chart process of DFM and DFA [5].

operates on the principle that the method of production influences the product design, and vice versa. It enables the adaptation of any product design to the current production infrastructure or the selection of low-cost methods that may be deemed a quick manufacturing upgrade [6]. This degree of collaboration, however, necessitates not only the development of novel tools but also a comprehensive re-evaluation of the integrated approach and the manufacturing processes of components and products. Sometimes, product designers conceive of intricate designs that provide formidable manufacturing challenges. In addition, designers frequently incorporate cutting-edge technology and techniques into their creations. Rarely are manufacturing methods that sophisticated, and modifying an established configuration may be prohibitively expensive. However, fast progress from idea to manufacture is only possible when the tools are well integrated [7].

A corporation may avoid, identify, measure, and reduce manufacturing inefficiency and waste at the design stage with DFMA. By developing the design in alignment with manufacturing capabilities, it is possible to circumvent the need for several revisions and design modifications that result in programme delays and cost escalation. By implementing the DFMA approach, development expenses are ultimately reduced due to the shortened time necessary to manufacture a unit, decreased assembly costs, reduction of process waste, and enhanced product dependability.

In addition to the manufacturing process, bringing a medical product to market necessitates adherence to rigorous regulatory criteria. To obtain US Food and Drug Administration (FDA) clearance for a novel medical device, meticulous planning and a delicate balance between technical advancement

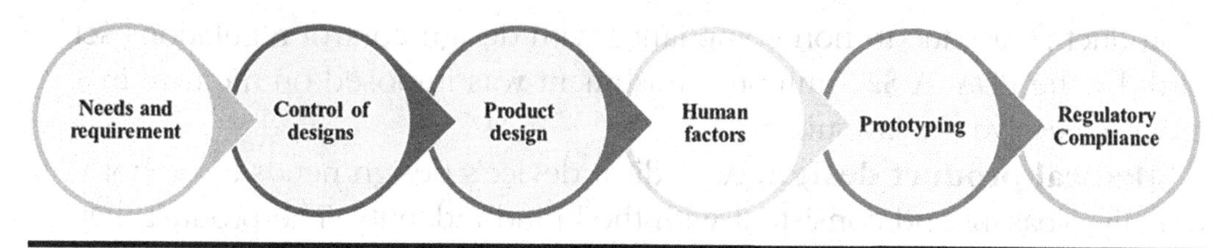

Figure 9.2 Overall aspects for design of healthcare products.

and product effectiveness are essential. Demands for time-to-market, engineering design cost, and future product cost limitations must also be mapped. For the protection of patients and users, the FDA has developed standards and recommendations that medical device manufacturers and designers must adhere to.

There are several aspects which could be considered before the design of a healthcare product (Figure 9.2). These aspects could help an entrepreneur gain insight into how a healthcare product is to be manufactured keeping in mind the economical and patient acceptance rates [8].

Needs and requirements of users: Recognising user wants and requirements is the initial and important phase in the development of medical devices. It is essential to comprehend the requirements of the end-user in order to design a device that is functional, user-friendly, and safe. It is imperative that medical device designers engage in comprehensive research in order to get knowledge pertaining to the user's requirements, inclinations, and constraints. Obtaining this information through user interviews, surveys, and observations is one approach. This methodology aids designers in developing a gadget that satisfies the user's requirements and effortlessly integrates it into their daily regimen. An instance of such a product is the continuous glucose monitoring system that was created by Medtronic for people with diabetes. Extensive research was conducted by Medtronic to produce a device that satisfied patients' requirements for continuous glucose monitoring without using finger sticks.

Control of designs: Design controls are procedures that guarantee the safety, efficacy, and dependability of the medical equipment. Design controls aid in the management of risks and guarantee that the product complies with regulatory standards. The utilisation of a risk management system, design validation and verification, and design transfer are all components of the design control procedure. These standards are particular to design controls mandated by the FDA, and designers of medical devices should be aware of them. For instance, Avanos Medical was compelled to initiate a product recall due to non-compliance with design control regulations set forth by the FDA. A $22 million punishment was imposed on the firm in order to resolve the lawsuit.

Medical product design: A medical device's design needs to be both visually pleasant and consistent with the brand's identity. The product that medical device designers should produce should be ergonomic, user-friendly, and simple to maintain. The formulation of medical products ought to incorporate an awareness of the device's lifecycle, encompassing its production, packing, and eventual disposal. An instance of exceptional design in the realm of medical products is the Philips Sonicare toothbrush. Aligned with the brand's aesthetic, the toothbrush is ergonomically built to sit comfortably in the user's hand and features a sleek, contemporary appearance. Additionally, the packaging of the gadget is created to have a reduced carbon impact in comparison to conventional packaging.

Considering human factors: Medical devices should be people's devices. Individuals engage with healthcare items in their capacity as professionals or patients [9]. It is imperative that these gadgets be developed with the end-users in mind. Human factors is the academic discipline concerned with the interactions and concerns that arise between humans and machines. Human factor concerns are equally as important as any other engineering, marketing, or manufacturing component when it comes to the design of healthcare products [10, 11]. For healthcare goods to be applied to a patient or operated by a medical professional in a safe manner, they must be developed with the user in mind. For example, a hearing aid could differ in several designs based on the requirement and ear designs (Figure 9.3). It could be highly possible that an individual could prefer one design over the other, hence specifically in medical devices, human factors consideration is of utmost importance.

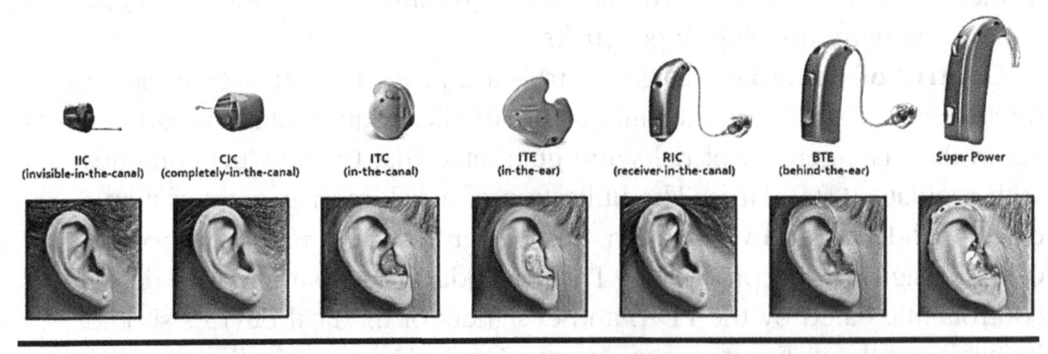

Figure 9.3 Different hearing aid designs for different purposes and ear designs [12].

Human factors encompass a vast array of statistically supported data concerning the physical and sensory limits of individuals. In order to achieve optimised product design, it is necessary to possess a thorough comprehension of the multifaceted user relationships that will occur with the product. Frequently, these elements consist of product characteristics that are directly influenced by human anatomy. The human brain is another key human factor component influencing the design of healthcare products. A complex combination of cerebral processes that impact human perception, response speed, understanding, and weariness has a significant bearing on every healthcare product. The rapidly expanding discipline of Graphical User Interface (GUI) focuses on the logical structure, operation flow, and visual arrangement of menus linked with medical equipment.

A full grasp of the product, its function, and the processes impacted by its use is necessary for GUI design. Menu item layout, icon design, and event sequencing are only a few of the many elements connected with GUIs. The incorporation of human elements is of the utmost importance when developing healthcare goods. Negligence of this nature, whether by omission or improper integration, may result in injury to the patient or legal ramifications. Human elements must never be undervalued while designing healthcare products.

Medical device prototyping: Prototyping is an essential component of the medical device design process. Before mass production, prototyping enables designers to evaluate the operation of the gadget, uncover design defects, and make any required improvements. Additionally, prototyping can aid in cost reduction and optimisation of the production process. A diverse range of prototype processes, such as 3D printing, machining, and injection moulding, have to be employed by medical device designers. Illustratively, 3D printing has the capability to produce complex components and pieces that are challenging to produce by conventional means.

Regulatory compliance: Ensuring regulatory compliance is of utmost importance in order to guarantee the safety and efficacy of the medical device for patients. It is essential that designers of medical devices have knowledge of the regulatory requirements of the countries in which the device will be sold. The FDA is responsible for regulating medical devices in the United States; prior to product marketing, firms are required to acquire FDA clearance. Additionally, designers of medical devices should take into account the Medical Device Regulation (MDR) of the European Union, which became operational in May 2021. In contrast to its predecessor, the Medical Device Directive (MDD), the MDR imposes more stringent criteria, including the need for further paperwork and testing.

On the other hand, industrial design is similar to DFM but focuses more on evaluating the appearance, practicability, functioning, user experience, and manufacturability of the medical equipment [13]. Incorporating industrial design across the whole process could guarantee a high-quality end result, reduced production expenses, and a streamlined product development cycle. There are a few steps for efficient industrial designing. The first step is the feasibility testing of ideas. At this stage, the test and analysis of the idea is carried out for its functionality and economic stability, and scalability. The second step is performing the competitive analysis. It becomes important to compare and analyse the idea for market needs. The third step is to analyse the market projection followed by the analysis of material to be used in the manufacturing. In the end, the entrepreneur could perform the overall cost structuring followed by the feedback to ensure end-user satisfaction.

All the above design steps may also be applied to a manufacturing facility that produces medical devices. A medical device manufacturing plant is a specialised site where pharmaceutical devices are conceived, manufactured, and quality-controlled to assure their efficacy and safety for use in the healthcare business in accordance with regulatory standards and guidelines. The initial phase of manufacturing plant setup is the design of the manufacturing plant layout. Medical apparatus manufacturing facilities are required to adhere to the regulations established by regulatory authorities such as the FDA. Orthopaedic implants, disposable medical devices, principal packaging materials, and other pharmaceutical goods are among the medical devices whose production facilities must be designed in accordance with Good Manufacturing Principles (GMP) and other regulatory standards.

When determining the optimal factory architecture for the medical device design facility, it is crucial to consider both the movement of personnel and materials. The significance of product segregation, process flows and production procedures, and the utilisation of categorised areas cannot be overstated. Effective guidance and support may result in accurate decision-making, therefore guaranteeing a seamless transition amongst audits conducted by regulatory agencies and customers. Figure 9.4 shows a typical layout of a medical manufacturing plant. It consists of common areas as observed in the production of any goods or items. The assembly section is attached to the washer room. In addition to this, the deburring room is adjacently attached to the assembly room so as to reduce the travel time. After this, the product is sent to the quality check (QC) lab and the items which fail the test are usually sent to rejection room.

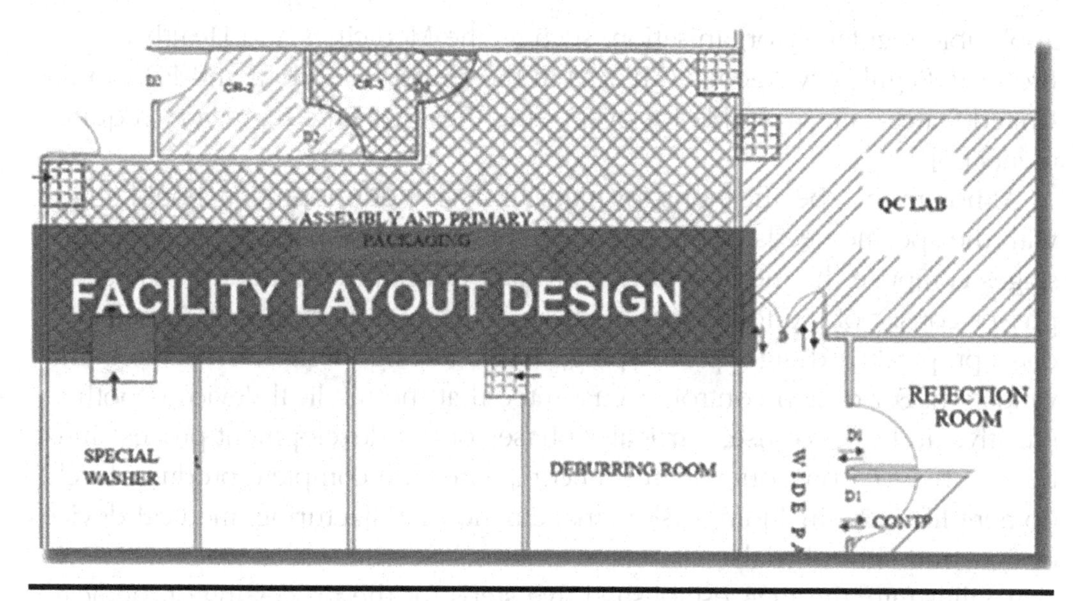

Figure 9.4 A typical layout of a medical device manufacturing plant [14].

The structure of a medical device manufacturing facility necessitates meticulous design and deliberation in a number of aspects, including regulatory compliance, workflow efficiency, and cleanliness. The process of creating a manufacturing plant layout for medical devices consists of the following steps:

1. Recognise the production procedure governing medical equipment.
2. Outline the manufacturing process's workflow.
3. Determine the regulatory compliance requirements
4. Develop an exhaustive strategy for the layout design.
5. Make best use of the area at hand to maximise efficiency.
6. Design with safety concerns as a top priority.
7. Architects and engineers should collaborate in the development of the final design.
8. Consistently assess and enhance the layout in order to guarantee adherence and effectiveness.

9.2 Stages of Processing Medical Devices

The precise product development process for medical devices varies by location, since distinct regulations and criteria must be adhered to in the European Union, United States, United Kingdom, and more areas. The applicable regulatory organisation, such as the Medicines and Healthcare Products Regulatory Agency (MHRA) in the United Kingdom and FDA in the United States, is responsible for overseeing these specific regulatory requirements [15].

Although the specifications for medical device development should align with the specific needs of the intended application area, some overarching stages comprise the medical device product development lifecycle. These phases consist of the following: medical device manufacturing, original concept, product design, prototyping, device testing, design verification, and validation. For design control to guarantee that the medical device is both effective and safe for use, particular phases of the development process must be adhered to. Consequently, this encompasses the complete product development lifecycle, including risk management, manufacturing, medical device design, and clinical trials.

To illustrate, the FDA established five steps for the processing of medical devices prior to their commercialisation. The aforementioned phases constitute their quality system regulation (QSR), which dictates "the processes, facilities, and controls utilised in the fabrication, assembly, packing, labelling, storage, installation, and maintenance of all completed products intended for human consumption."

FDA phases consist of:

1. Analysis of opportunities and risks from the outset.
2. Conceptualisation and feasibility examination of formulation.
3. Design and development, including validation and verification to guarantee that the design output corresponds to the design input specifications.
4. Final confirmation and preparation for the product launch.
5. Product introduction and post-launch evaluation.

The process of developing and obtaining clearance for a novel medical device is protracted, potentially spanning many months to years, dependent upon the specific nature of the device under development [16, 17]. Three to seven years are usually required to bring a technology from idea to approval, according to studies. While it may appear to be an extended period, this encompasses the complete lifespan of the gadget, which includes research, development, and testing. The FDA establishes timeframes for obtaining permission to introduce several categories of devices to the market [18]:

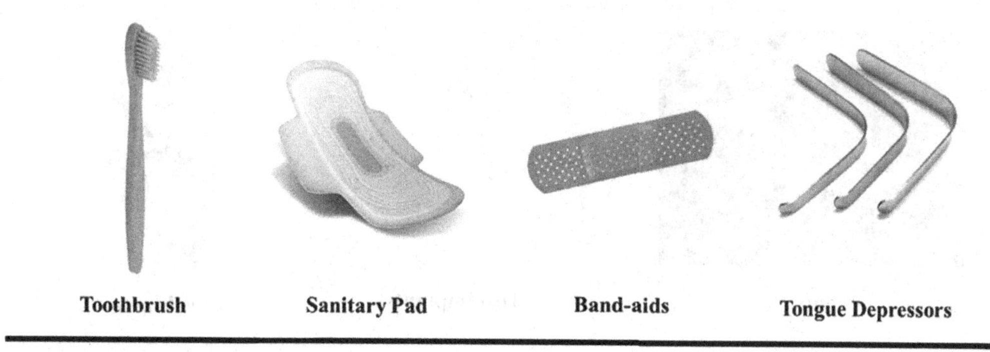

Figure 9.5 Examples of class 1 medical products.

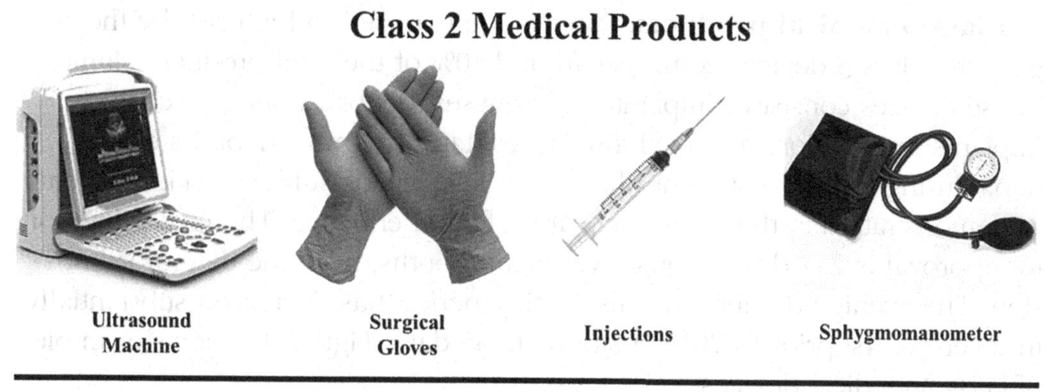

Figure 9.6 Examples of class 2 medical products.

Class 1 medical products: These consist of gadgets such as electric toothbrushes, tongue depressors, and oxygen masks, which obtain FDA approval the quickest. Self-registration of the FDA for the majority of these non-invasive class 1 devices may be possible in as little as one week. Figure 9.5 shows common examples of class 1 medical products.

Class 2 medical products: These consist of products like syringes, catheters, and contact lenses, which carry a moderate level of risk. This category includes 43% of all devices; thus, a maker must demonstrate the safety and effectiveness of the item by a considerable comparison to an already-approved device. Although the FDA guarantees acceptance of an application for approval within 60 days, clearance is typically granted after 177 days (almost six months), with just 19% of submissions undergoing the clearance process within three months. Nevertheless, clearance times may differ based on the specific device type that is submitted. The average time to approval for anaesthesiology devices is 245 days, whereas the average time for toxicological devices is 163 days. Figure 9.6 shows some common examples of class 2 medical devices.

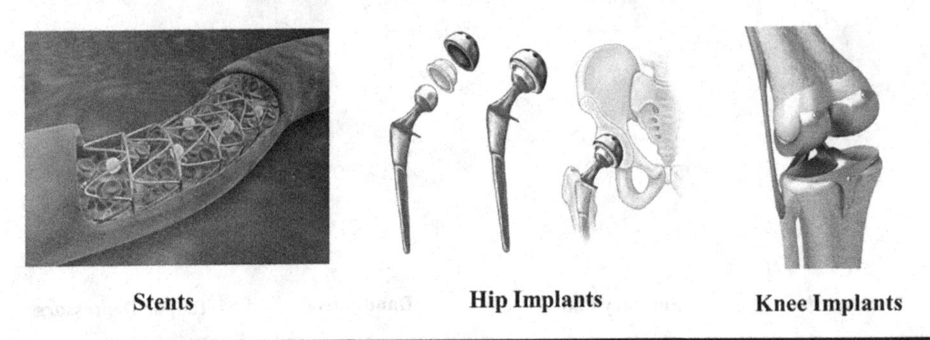

Figure 9.7 Examples of class 3 medical products.

Class 3 medical products: Patients are exposed to high risk by these gadgets. Class 3 devices comprise around 10% of the total product volume. These devices consist of implanted prostheses, defibrillators, and cochlear implants. In addition, approval also necessitates the most rigorous evaluations during the development phase. The FDA needs sufficient scientific data to demonstrate that these products are safe and effective. The mean duration for approval is 243 days, or just over eight months, from the date of submission. The average duration of this waiting period has decreased substantially in recent years; prior to 2010, it stood at 345 days. Figure 9.7 shows examples of class 3 medical devices.

Costs connected with the development of a medical device will vary in detail based on the specific item under development. The relative development expenses of a tongue depressor and a defibrillator will reflect the obvious variations in their levels of complexity. In addition, cost savings may not be achieved by skipping any of the iterative development, design, or testing phases. Prior to introducing a product to the market, a device maker must demonstrate several aspects, including technology, industrial design, safety, usability, proof of concept, manufacturability, and packaging designs.

9.3 Challenges and Future Directions of Medical Device Manufacturing

The manufacture of medical devices is one of the most dynamic industries due to the fact that it is always evolving to satisfy the shifting demands of consumers. The regulatory environment in which medical device makers operate is complicated, with ever-changing norms and expectations.

Manufacturers face a substantial obstacle in the form of harmonising global standards and staying abreast of evolving laws [19]. Nonetheless, it is critical to be informed as well. Incorporating compliance knowledge, being up to date on regulatory changes, and sustaining resilient quality management systems are critical for organisations to effectively traverse this complex environment. All of it, however, comes at a price. Furthermore, for continued success, organisations must consistently strike a delicate equilibrium among quality adherence, expenditures, and output.

Some major issues could be cybersecurity and data privacy issues. Patients' medical records are among the most private items in their possession. Nevertheless, it is an undeniable reality that the interconnected world of today facilitates universal access to them with a mere few mouse clicks. Everything is interconnected, from mobile phones to wearable medical equipment to hospital data, in order to improve patient care. The heightened level of connection has further complicated the task of efficiently protecting patients' cybersecurity and data privacy. Medical gadgets that are connected are susceptible to cyber assaults. This may involve data breaches, illegal access, or other forms of manipulation. To secure patient data and guarantee the safety and integrity of connected medical equipment, manufacturers must adopt sophisticated cybersecurity protections, including authentication and encryption procedures.

A growing concern is personalised medicine and patient-specific device manufacturing. The emergence of customised medicine and patient-specific technologies poses medical device manufacturers with both prospects and obstacles. Technological advancements, like additive manufacturing and 3D printing, allow the fabrication of personalised medical equipment that is specifically designed for each patient. Personalised medicine has the capacity to fundamentally transform the healthcare industry through the provision of customised treatments and treatment options. But considering the manufacturer's perspective, this presents an even greater obstacle.

During manufacturing, environmental issues pose several risks. Thereby, an escalating number of manufacturers are adopting recycling programmes and streamlining manufacturing processes to save waste while using environmentally friendly products. In the medical device business, the notion of a circular economy, in which materials are reused or used, is gaining support. Nonetheless, everything is implementable to a certain degree. Due to the sensitivity of medical applications with regard to potential cross-contamination and toxicity, certain materials cannot be recycled or repurposed.

References

1. Design for manufacturing - medical device | Johari Digital Healthcare Ltd n.d. https://www.joharidigital.com/design-for-manufacturing/ (accessed December 11, 2023).
2. Battaïa O, Dolgui A, Heragu SS, Meerkov SM, Tiwari MK. Design for manufacturing and assembly/disassembly: joint design of products and production systems. *Int J Prod Res* 2018;56:7181–9. https://doi.org/10.1080/00207543.2018.1549795.
3. Design for manufacturing medical devices | Kapstone Medical n.d. https://www.kapstonemedical.com/services/manufacturing-solutions/design-for-manufacturing (accessed December 11, 2023).
4. Xiao A, Seepersad CC, Allen JK, Rosen DW, Mistree F. Design for manufacturing: application of collaborative multidisciplinary decision-making methodology. *Eng Optim* 2007;39:429–51. https://doi.org/10.1080/03052150701213104.
5. Yin C, Hou S. Application design for manufacture and assembly to improving product design and development performance. *2019 ASABE Annu Int Meet* 2019:1. https://doi.org/10.13031/AIM.201900289.
6. What is DFM – design for manufacture? | Arrotek n.d. https://arrotek.com/what-is-dfm-design-for-manufacture/ (accessed December 11, 2023).
7. Wasim M, Vaz Serra P, Ngo TD. Design for manufacturing and assembly for sustainable, quick and cost-effective prefabricated construction – a review. *Int J Constr Manag* 2022;22:3014–22. https://doi.org/10.1080/15623599.2020.1837720.
8. What are the 5 key aspects of medical device design? n.d. https://www.cambridge-dt.com/what-are-the-5-key-aspects-of-medical-device-design/ (accessed December 11, 2023).
9. Healthcare product design: The chemistry of applying the human element n.d. https://idsys.com/healthcare-product-design/ (accessed December 11, 2023).
10. Gupta S, Jayaraman R, Sidhu SS, Malviya A, Chatterjee S, Chhikara K, et al. Diabot: development of a diabetic foot pressure tracking device 2023;6:32–47. https://doi.org/10.3390/J6010003.
11. Gupta S, Malviya A, Chatterjee S, Chanda A. Development of a portable device for surface traction characterization at the shoe-floor interface. *Surfaces* 2022;5:504–20. https://doi.org/10.3390/SURFACES5040036.
12. Common types and styles of hearing aids - perfect hearing n.d. https://perfect-hearing.my/types-styles-hearing-aids/ (accessed December 11, 2023).
13. Industrial design for medical devices: 6 steps to develop a successful device n.d. https://www.joharidigital.com/medical-device-industrial-design/ (accessed December 11, 2023).
14. Manufacturing plant layout design for medical device (Factory layout design consultant) | Operon strategist n.d. https://operonstrategist.com/services/turn-key-project/manufacturing-facility/ (accessed December 11, 2023).
15. Medical devices – Angela N Johnson n.d. https://angelanjohnson.com/medicaldevices/ (accessed December 11, 2023).

16. Sucher JF, Jones SL, Montoya ID. An overview of FDA regulatory requirements for new medical devices. *Expert Opin Med Diagn* 2009;3:5–11. https://doi.org/10.1517/17530050802644673.
17. Mooghali M, Ross JS, Kadakia KT, Dhruva SS. Characterization of US food and drug administration class I recalls from 2018 to 2022 for moderate- and high-risk medical devices: a cross-sectional study. *Med Devices Evid Res* 2023;16:111–22. https://doi.org/10.2147/MDER.S412802.
18. What is the medical device development process? (Includes Stages) - TWI n.d. https://www.twi-global.com/technical-knowledge/faqs/what-is-the-medical-device-development-process (accessed December 11, 2023).
19. Your ultimate guide to medical device manufacturing | Rapid direct n.d. https://www.rapiddirect.com/blog/medical-device-manufacturing/ (accessed December 11, 2023).

PROTOTYPING

Chapter 10

Minimum Viable Product, Wireframing, and Introduction to App Development

10.1 Minimum Viable Product

The effectiveness, accessibility, and dependability of healthcare services have been enhanced in tandem with technological progress [1]. The emergence of online platforms and applications has brought about a significant transformation in the healthcare industry through the provision of expedient and easy means to interact with medical professionals, track one's health, retrieve electronic data, and undertake other tasks [2].

Numerous healthcare entrepreneurs are motivated to develop their own devices and applications in pursuit of novel commercial prospects, drawing inspiration from the success of other healthcare software companies [3]. However, it is risky to invest in a new entrepreneurial concept, particularly one that involves technology [4]. Entrepreneurs may begin by developing a minimal viable product (MVP), which enables them to validate their concept while mitigating potential hazards. It is the initial phase of digital healthcare product development that permits real-world audience usability assessment [2]. Even established healthcare companies may utilise MVP development to evaluate the efficacy of their technological solutions, which are designed to optimise patient care and expedite internal operations. Hence, understanding MVP is important before developing the actual working prototype [5].

DOI: 10.4324/9781003475309-14

MVPs are products that have a sufficient number of functionalities to entice early adopters and verify product concepts at an early stage of the product development life cycle [6]. The MVP may assist the product team in sectors such as software in receiving customer input as rapidly as possible so that the product can be improved via iteration [7]. Agile development is predicated on verifying and iterating products in response to user feedback; hence, the MVP is of paramount importance [2]. Most healthcare entrepreneurial ventures could fail due to lack of market needs for their services or products. Hence, upfront planning could turn out to reduce these failures [8]. For instance, the MVP iteration of a remote patient monitoring programme need not be an all-encompassing extension. Also, there has been a significant increase in the number of healthcare applications in the last few years [9].

There are different types of healthcare MVPs that could be used by entrepreneurs. To discuss a few, in the software domain, these could be namely healthcare and fitness, telemedicine, reminders, remote monitoring, and doctor on-demand applications.

Healthcare and fitness applications: Especially after the COVID-19 pandemic catastrophe, individuals are becoming increasingly concerned with their health and fitness. Consequently, the prevalence of health and fitness applications has surged [3]. These intuitive applications enable users to establish fitness objectives such as daily step count or calorie consumption. Developing an MVP for health and fitness applications may be very straightforward in comparison to other healthcare products; nevertheless, the intricacy may increase when the applications incorporate more sophisticated functions. Figure 10.1 shows a typical graphic of a healthcare and fitness application.

Telemedicine: A landmark in healthcare services has been reached with the advent of telemedicine applications. Exceptional features of these technologies provide instantaneous communication between patients and physicians [11]. Patients are able to communicate with physicians of various disciplines using telemedicine or telehealth technologies. Regarding treatments, patients may communicate with physicians by SMS, video call, phone call, and so on. These solutions also facilitate the effective and straightforward scheduling of appointments for patients. Figure 10.2 shows a generic telemedicine application which shows the vital statistics of a patient.

Reminder: Adherence to medication schedules is crucial and efficacious in the treatment of illnesses. However, due to the hectic nature of life, it is easy to overlook one or more medication doses [12]. In such circumstances,

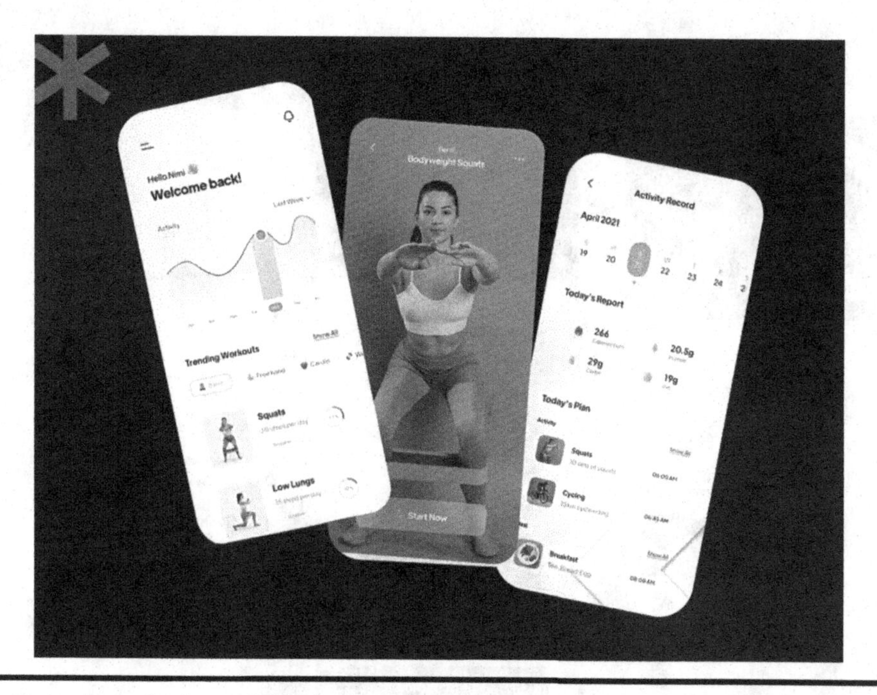

Figure 10.1 A typical graphic of a healthcare and fitness application [10].

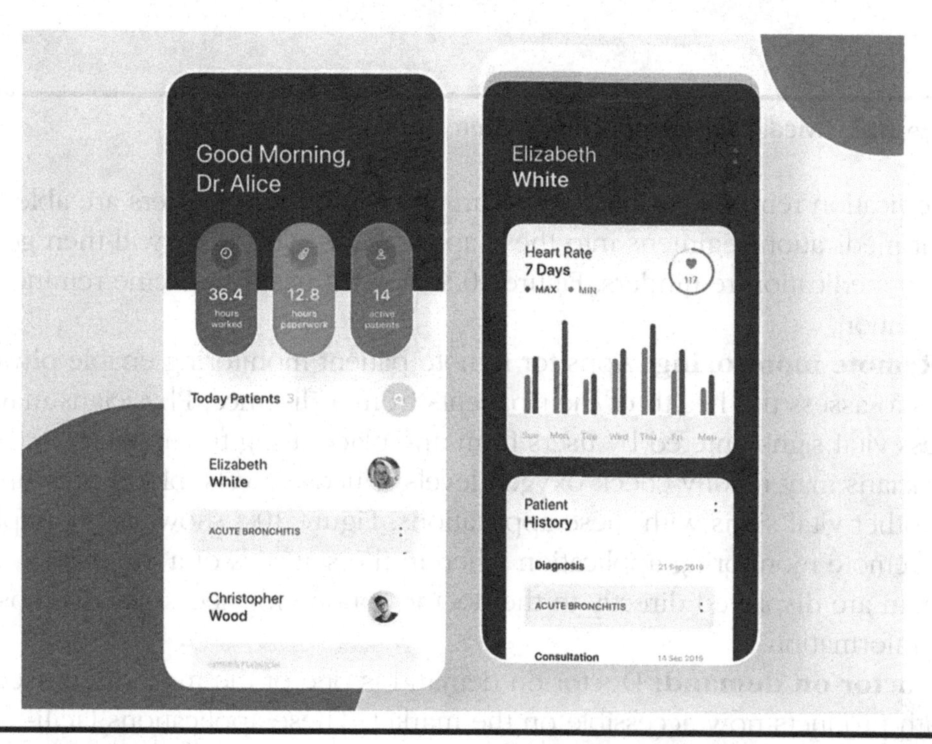

Figure 10.2 A generic telemedicine application which shows the vital statistics of a patient [10].

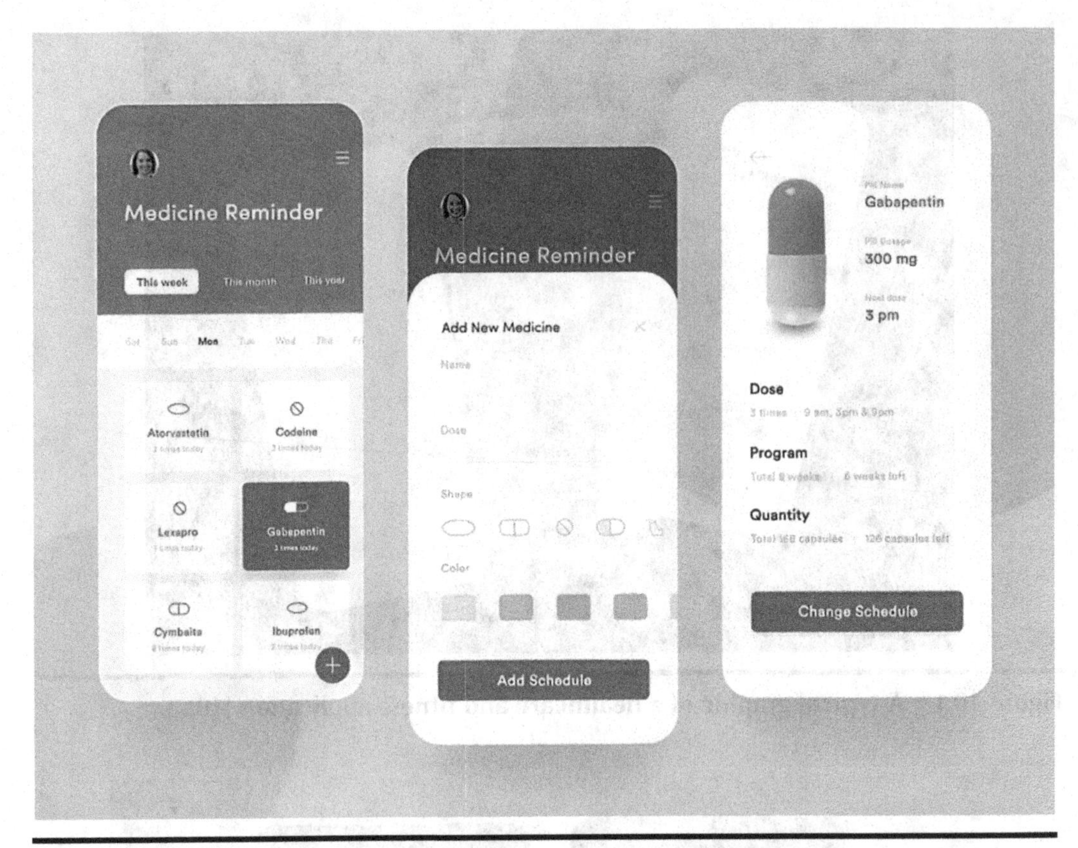

Figure 10.3 Medicine reminder application [10].

a medication reminder application is an absolute lifesaver. Users are able to input medication regimens into these applications. The users will then get timely medication reminders. Figure 10.3 shows a usual medicine reminder application.

Remote monitoring: Apps for remote patient monitoring enable physicians to assess the health of their patients from a distance. Physicians may access vital signs entered by users from any place using the application [13]. Physicians may readily check oxygen levels, glucose levels, blood pressure, and other vital signs with these applications. Figure 10.4 shows an example of a remote monitoring application wherein the statistics of a pregnant woman are displayed directly to the doctor's phone for necessary diagnosis and information.

Doctor on demand: Doctor on demand is one of the most often-used health products now accessible on the market. These applications facilitate the provision of patient care by physicians via video conversations. Physicians have the ability to communicate with patients from any location

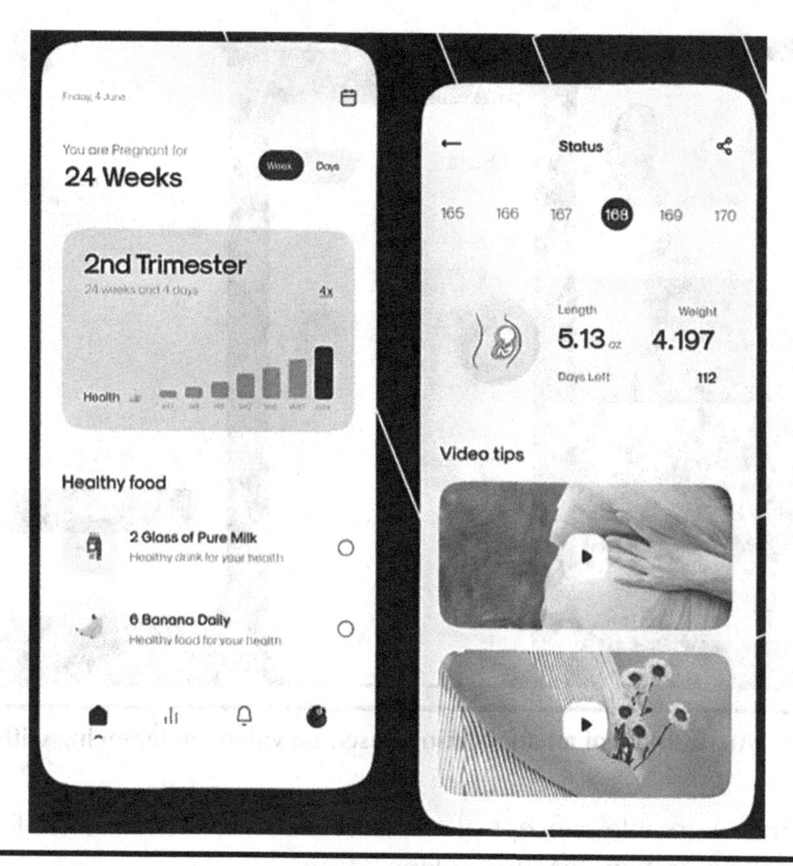

Figure 10.4 An example of a remote monitoring application tracking a pregnant woman's health [10].

on the planet. They are able to conveniently observe patients and administer medical care. Figure 10.5 shows an example of an application based on video conferencing with a doctor.

In order to develop the MVP, certain steps may be followed. Nevertheless, the development process is marginally more manageable with an MVP. MVP development entails the same sequence of activities as full-scale development, commencing with the conception phase and concluding with the release of your healthcare software product or application. The stages involved in developing an MVP for a healthcare application are detailed below to facilitate comprehension [2].

Planning: Determining the health sector for which an entrepreneur intends to create an application or software solution is the initial step. Determine if they intend to create an application for telemedicine, health and fitness, health monitoring, or an entirely different form of digital product. This phase entails the transformation of undeveloped software product

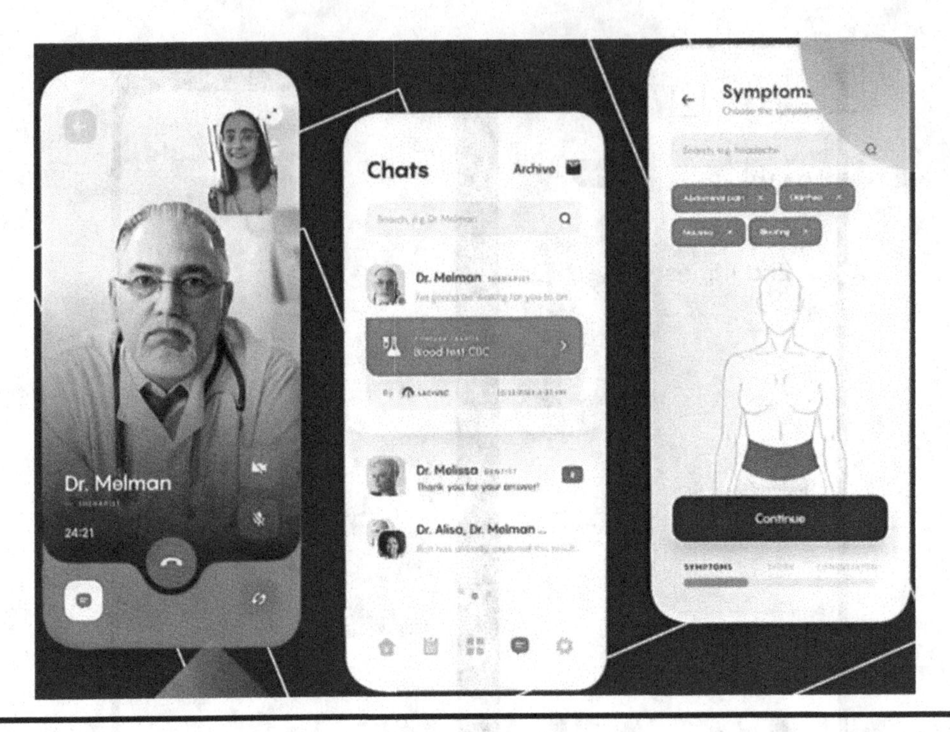

Figure 10.5 An example of an application based on video conferencing with a doctor [10].

concepts into workable concept plans, as well as the formulation of an implementation strategy for those plans.

Design: User interface/user experience (UI/UX) design constitutes the following phase in MVP development. UI/UX designers may develop the user interface for the healthcare software product at this phase in order to construct an interactive and feature-rich solution that appropriately addresses the requirements of the end-users.

Development: During this phase, the device's, or the software or application's functionalities will be constructed. MVP development will be conducted utilising cutting-edge technology and an agile software development lifecycle.

Testing: The subsequent critical phase in the creation of a healthcare MVP is quality assurance. This phase encompasses the examination of the product or application product to verify its functionality and quality, as well as its viability for commercial release.

Launch: Upon completion of development and testing, the MVP for healthcare may be ready for release. The product in question will be accessible and utilised by real consumers. Entrepreneurs may monitor their input and incorporate improved functionality to enhance their healthcare product.

10.2 Business Model Canvas

As discussed, MVP is a strategic technique to showcase the healthcare product or a service in an effective manner, other techniques such as developing a business model canvas could also help understanding the workflow of an entrepreneurial venture [14]. The business model canvas (BMC) is a strategic management tool to rapidly and effectively establish and express a business idea or concept [15, 16]. It is a one-page document that works through the key parts of a business or product, outlining a concept in a consistent way. Figure 10.6 shows a typical template of a business model canvas which could be used by entrepreneurs filling out the segments [17].

The following describes each segment of a BMC. The following explanation could be used by an entrepreneur to enter the details into the canvas.

Value proposition: Value propositions are fundamental to all products and businesses. They constitute the foundational principles underlying the value transfer that occurs between an enterprise and its clientele [18]. Generally, customers exchange value for monetary compensation when the entrepreneurial ventures resolve an issue or alleviate their suffering.

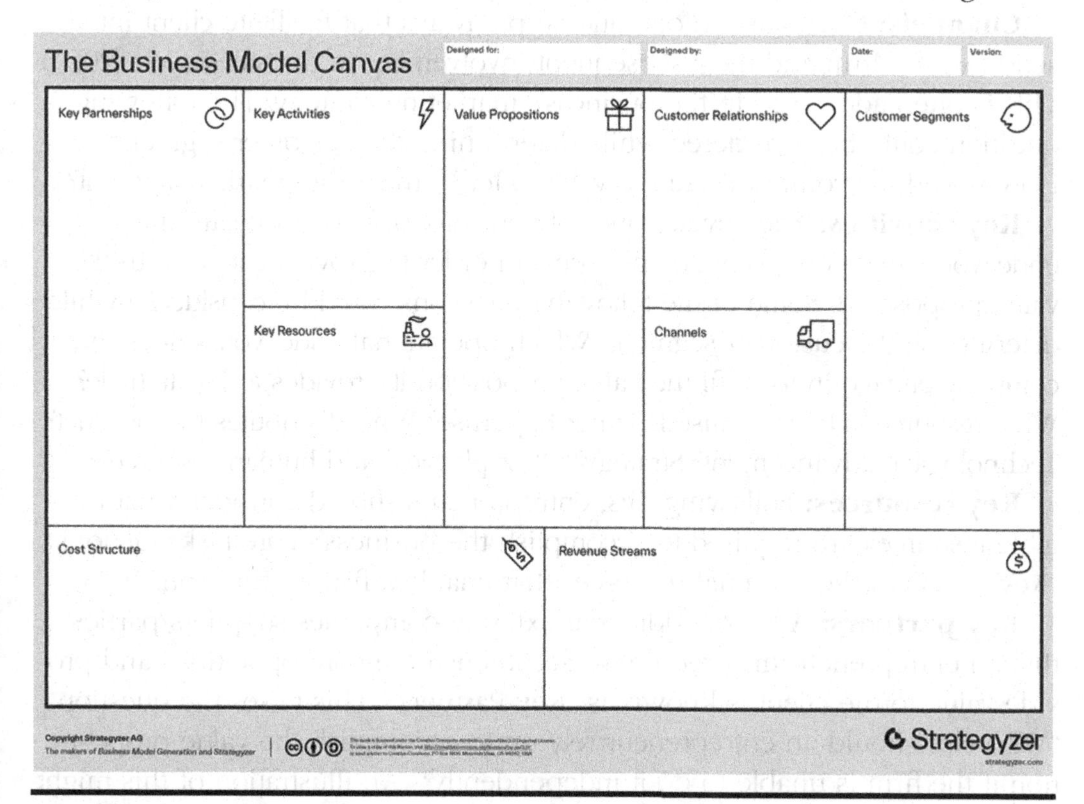

Figure 10.6 A typical template for business model canvas generation [17].

Entrepreneurs should think what problems they are solving, why would their customers want this problem to be solved, what could be the motivating factor for that problem.

Customer segments: Client segmentation is the partitioning of a customer base into distinct groups of individuals who share commonalities, including but not limited to age, gender, hobbies, and purchasing patterns. Some of the following questions could be considered while determining the customer segment. For whom are the entrepreneurs addressing the issue? Who could place importance on value proposition? Is the value offer attractive to both men and women? Is it appealing to adolescents or young adults aged 20 to 30? Which attributes define the individuals who are in pursuit of the value proposition?

Customer relationships: Customer relationships refer to the dynamic between an organisation and its clientele [19]. Developing a user journey map depicting the user's interactions with the organisation is an exceptionally beneficial initial action. This facilitates the clarification of the points of interaction and ways of communication between the entrepreneur and their consumer. Additionally, this could aid in the initial delineation of the business activities and the detection of potential areas for automation.

Channels: Channels are operational pathways that facilitate client interaction with a firm and their subsequent involvement in the sales cycle [20]. This is often addressed in the business's marketing strategy. The following question could be considered while determining the customer segment. How would we communicate the value offer to the target market segment?

Key activities: The key activities of your product or service are the endeavours that your company performs in order to provide clients with the value proposition. Some of the following questions could be considered while determining the customer segment. Which operational endeavours does the company engage in to fulfil the value proposition it provides to its clientele? What resource is being utilised? Time? Expertise? Who distributes the product? Technological advancement? Strategy? Offer physical and human resources?

Key resources: Following this, entrepreneurs should consider what tangible resources are required to accomplish the business's core tasks (actions). "Key" refers to the essential resources that enable a firm to function.

Key partners: A list of additional external companies/suppliers/parties that an entrepreneur may require to accomplish the core operations and provide value to the client is known as "Key Partners." This raises the question, "Who else should an entrepreneur rely on to accomplish the value proposition if the firm is unable to do it independently?" An illustration of this might be "if an entrepreneur operates a grocery shop and sell to consumers, they could require fresh bread from a local baker."

Cost structures: The company's cost structure consists of the financial expenses associated with conducting business. Some of the following questions could be considered while determining the customer segment. How much does it cost to accomplish the essential operations of my business? How much do the most valuable assets and relationships cost? What is the expenditure associated with developing a value proposition for the clientele or users? Are there supplementary expenditures associated with operating a business? Legal? Insurance? What is the company's operating expense? Additionally, it is critical to provide a monetary value to the time as a cost. What is the total cost of employing staff?

Revenue streams: Revenue streams refer to the mechanisms via which a company generates financial benefit by converting its value proposition or resolving a customer's issue. Additionally, it is critical to comprehend how to price the company in accordance with the customer's pain of purchase, in return for the difficulty of resolving the issue.

In the above example (Figure 10.7), a sample business model canvas for a healthcare entrepreneur is shown for a new hospital. Usually, for a new hospital, the key partners are listed as medical supply companies, other

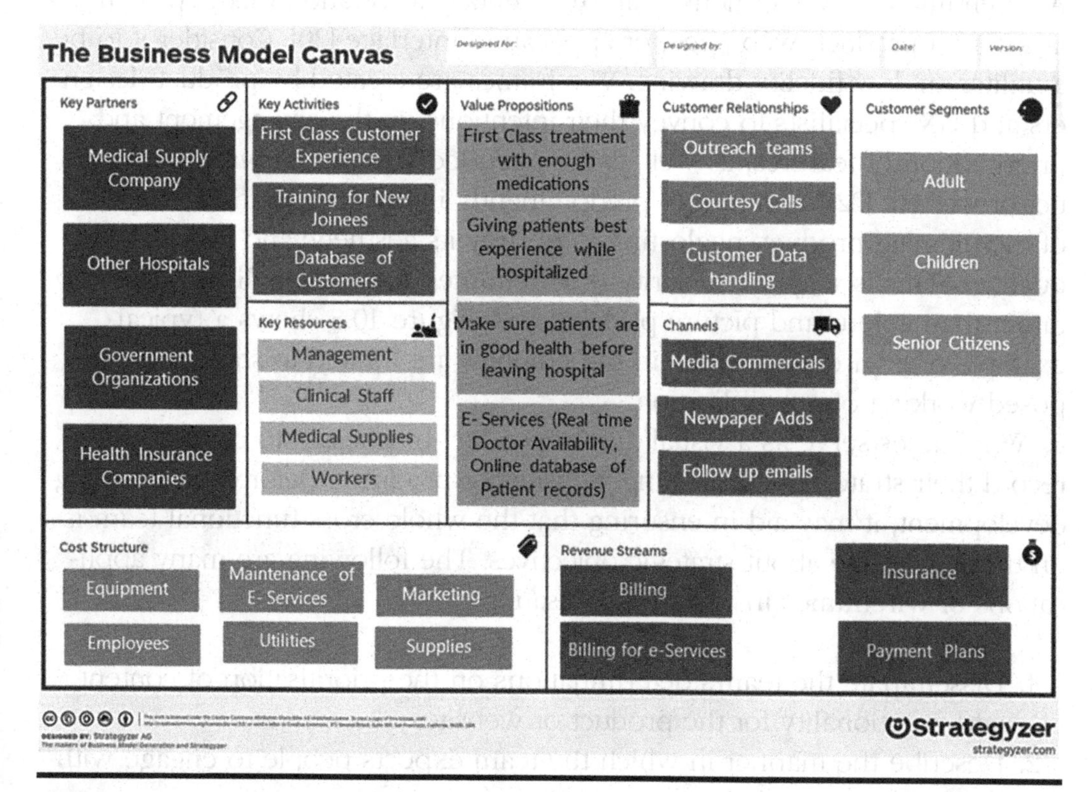

Figure 10.7 An example of a filled business model canvas for opening a new hospital [21].

hospitals, government organisations, or health insurance companies. Key activities include first class customer experience, training for new joinees, and database for customers or patients. Its key resources include the administration staff, clinical staff, medical supplies, and sanitation workers. The key value propositions for a new hospital could include first class treatment with appropriate medicines, giving the best experience to patients, making sure that patients are in good health, and availability of online services. Customer relationships include outreach teams, courtesy calls, and customer data handling. On the other hand, customer segments include adults, children, and senior citizens. There could be several channels such as media commercials, newspaper advertisements, or follow-up emails. The cost structure could include equipment, employees, maintenance of online services, marketing, supplies, and utilities. Finally, revenue streams include billing, insurance, payment packages, and billing for e-services.

10.3 Wireframing

A wireframe is a two-dimensional, rudimentary illustration that depicts the layout of a product, web page, or application interface [20]. Consider it to be a utilitarian, low-fidelity drawing. Wireframes are created by product designers and UX specialists to convey their intentions on the arrangement and prioritisation of features, as well as the intended UX and UI with the product or website [22]. Wireframes predominantly illustrate operational aspects of the ultimate product, neglecting to portray its aesthetic and functional qualities. This is why the majority of wireframes appear simplistic: greyscale rather than colour and picture placeholders. Figure 10.8 shows a typical example of a wireframe wherein the graphics are limited to show the proposed working of an application.

Wireframes serve as a visual aid for product businesses to convey and record their strategies for creating websites or products. Before commencing development, it may aid in ensuring that the whole cross-functional team is on the same page about strategic objectives. The following are many applications of wireframes inside an organisation:

1. Disseminate the team's determinations on the prioritisation of content and functionality for the product or website.
2. Describe the manner in which the team expects people to engage with the product or website.

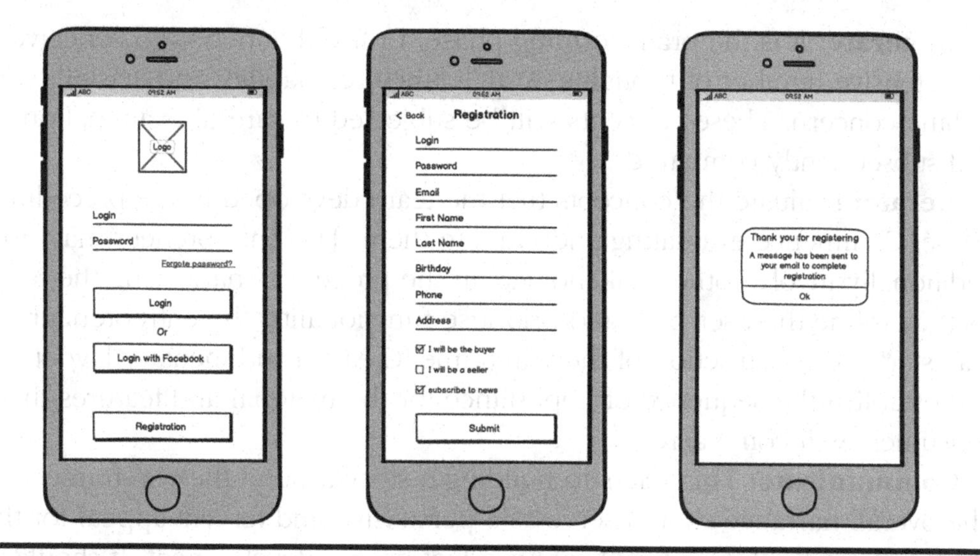

Figure 10.8 A typical example of a wireframe wherein the graphics are limited to show the proposed working of an application [27].

3. Document and convey the interconnections between the app's or website's various components.
4. Develop standardised methods for presenting certain material throughout the website.
5. Provide company-wide stakeholders with clarity on the product team's intended UX and product development.

Wireframes are often drafted by product designers, including UX and UI designers, at several businesses. Early in the product planning process, they accomplish this in collaboration with the product management team. Once the product team has made a decision on the prioritisation of features and the desired UX, the design team will endeavour to visually represent this information through the wireframe [23]. The procedure will be iterative in nature. The cross-functional team will collaborate with stakeholders from many departments, including design and product, to enhance the wireframe until it aligns with the organisation's strategic goals for the application. To generate a wireframe, the following steps could be used.

Articulate: Ensure that the team is cognisant of the issue they are attempting to resolve. Replying to inquiries of this nature serves as a solid foundation. The following questions could be used: Who will be the end-user? What are their objectives? How do they anticipate attempting to accomplish those objectives?

Generate: It is the brainstorming phase. Utilise this period to conceive of an extensive number of thoughts. At this juncture, viability and feasibility are of little concern. These concepts will be subjected to careful examination and subsequently compared.

Iterate: Evaluate the concepts that the team developed in the preceding phase. Commence evaluating and ranking them. The entrepreneur may now additionally involve other stakeholders in the process of narrowing the list. As they refine the user path and prioritise functionality, the entrepreneur may start the construction of the wireframe. Create a rudimentary layout and establish the sequence of importance for the material and features that consumers will come across.

Communicate: This refers to building a story around the wireframe. The overall narrative should serve as a persuasive and fervent appeal for the proposed approach to a healthcare product or app development. It should explain to the user persona why this product will address tangible issues and why they will consider the solution valuable enough to justify the cost. This narrative may be utilised, for instance, to introduce the wireframe to the executive team. Additionally, the visual overview of your wireframe must be translated into more comprehensive instructions for the development team responsible for constructing the final product.

Validate: The entrepreneur's team has dedicated time to conducting research, engaging in strategic brainstorming, and constructing the wireframe thus far. However, the team has not yet begun constructing a real product or developing the application code. Before devoting that amount of time and money on it, there are a few things that the entrepreneur may verify. The entrepreneur may first determine whether the suggested product is feasible and deserving of further development. It may entail, for instance, providing a representative sample of the team's target users with the wireframe in order to determine their degree of interest. Second, confirm with the technical team that they are capable of constructing a product according to the specifications in the developed wireframe.

The example in Figure 10.9 is mentioned as a wireframe for the development of a healthcare, wellness, and fitness application. It should be noted that the example just contains the sketched outlines of the proposed application and not the actual application. This wireframing technique could be used to showcase the product to the testers as a viability of the service. Also, it could be used to pitch to investors. In the Figure 10.9, the wireframe starts with a signup page followed by a login page. The application then asks to enter the personal details such as name, gender, phone number, age, weight,

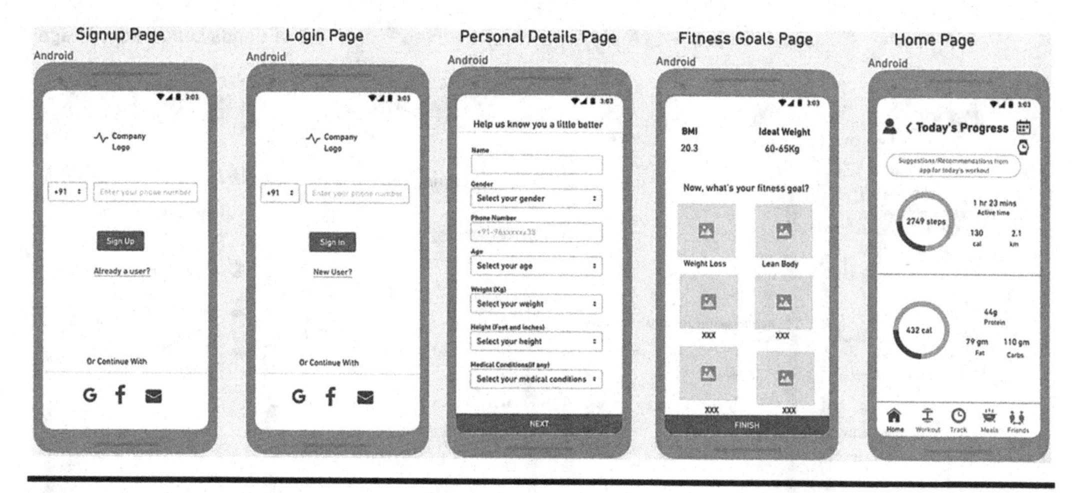

Figure 10.9 Wireframe of a healthcare, fitness, and wellness application: signup and login page [20].

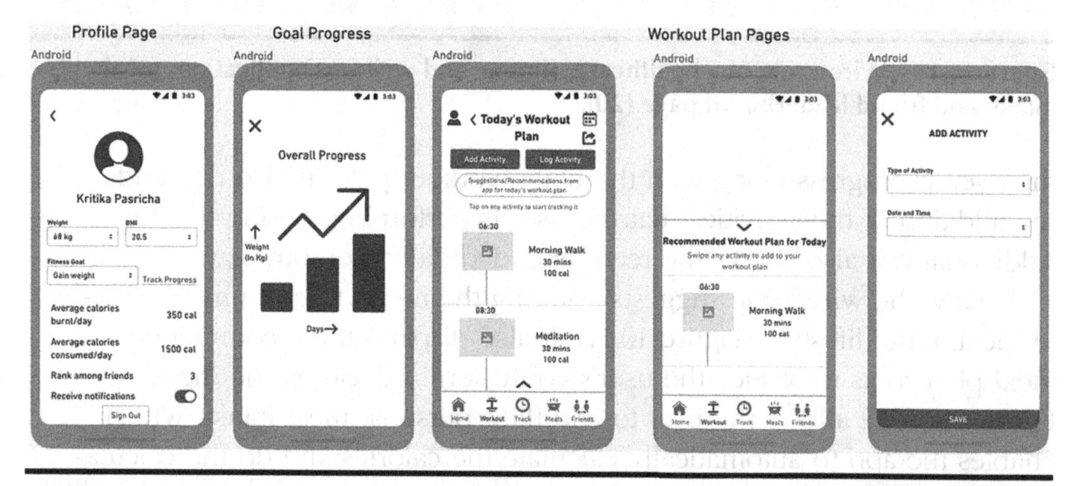

Figure 10.10 Wireframe of a healthcare, fitness, and wellness application: profiles and goals page [20].

height, and medical conditions experienced in the past. Furthermore, the next page includes fitness goals which include weight loss or lean body, etc. After the details are entered, the user is finally directed to the home page which contains the progress statistics. The statistics include total number of steps per day, and the calories burnt.

Furthermore, there is a separate page for profile which asks the weight and shows the body mass index (BMI) and fitness goals (Figure 10.10). It further shows the average calories that were burnt for the day. On the other hand, it also shows the calories consumed during the day. On the bottom it displays the sign-out button. In the following pages, the wireframe shows

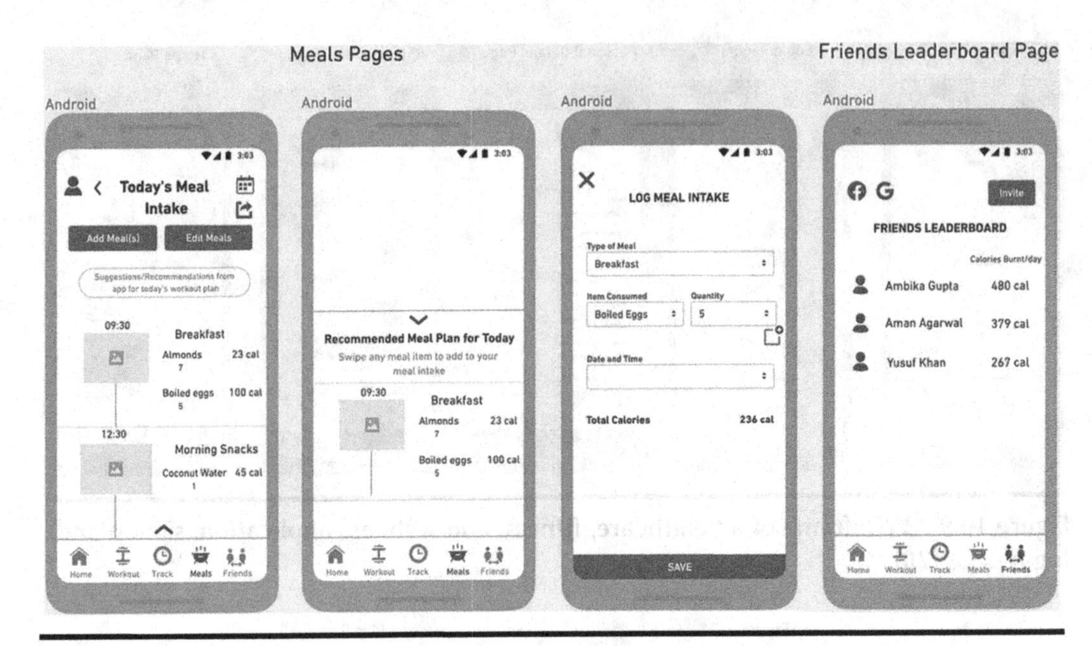

Figure 10.11 Wireframe of a healthcare, fitness, and wellness application: meal intake and friend leaderboard page [20].

the overall progress along with the daily workout plan. It also demands the addition of the activities that the user is performing during the day. Additionally, it also shows the recommended daily workout plan.

Finally, the wireframe suggests entering the meals to consume for better health and lifestyle (Figure 10.11). It also displays the recommended meal plan so as to reduce the user's confusion in deciding the meals. Furthermore, it allows the user to add their personal meal intake which enables the app to automatically calculate the calories and do the calculations. In the end, the friend's leaderboard is displayed which is anticipated to keep the users motivated by looking their friend's progress according to their fitness goals.

10.4 Introduction to Application Development

Wireframes may be hand-sketched or drawn on a computer. Although sketched wireframes work, a real-world working prototype of an application could be a better option to pitch the idea [24]. To achieve this, an entrepreneur could consider several freely available application prototyping services to generate their wireframes. As these services do not require expertise in computer programming, these freely available services are highly used to graphically pitch the ideas to investors or users. There are several wirefram-ing websites which could be used. However, "Figma" [25] is a popularly used application prototype service.

By utilising the web-based design application Figma, users are able to produce collaborative, interactive designs. Figma enables users to generate or modify graphical components such as pictures, icons, and drawings. The application features a contemporary and user-friendly design, allowing for effortless item selection using the pointer and modification via the selection of an icon from the toolbar located at the top of the screen. When alter-ing an image, modifications performed in one section of the document are promptly reflected in the other sections, relieving designers of the concern of preserving changes or risking work loss. Designers may find Figma's capa-bility to publish files optimised for mobile device viewing to be especially advantageous. Entrepreneurs could follow the below steps to get started with Figma and create their application wireframes.

Creating account on Figma: The designer may first visit figma.com and register to get started with the sketch development. The designer could also follow a quick guided tour to get started.

Interface: The tools necessary for the creation of designs are accessible from the top left menu. Look around and investigate the following options:

Primary menu
Move tools
Region-specific tools
Forming tools
Drawing implements
Manual text tool
Include a remark

Page 1, Assets/Strategies: Each element that is generated may be located in the menu on the left. Panel inspection: In the menu on the right are the val-ues of the properties associated with your work. It is evident that there are three distinct tabs: design, prototype, and inspect (Figure 10.12).

Canvas: Upon generating a fresh design file, the canvas will be rendered. This is the whole of the Figma workspace.

Frames: Figma's artboards are called frames. Select F from the keyboard or navigate to the Tools menu, where the user may click the Frame icon to gen-erate a frame. Choose from a diverse selection of presets: smartphone, tablet, desktop, watch, presentation, paper, Figma community, or social media.

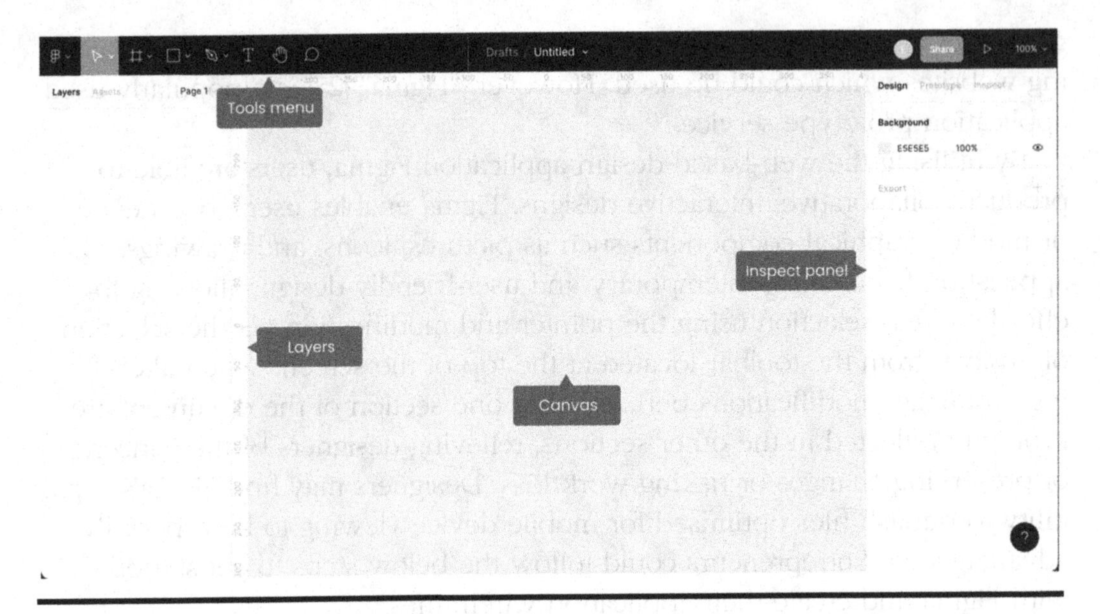

Figure 10.12 Figma layout [26].

Vector shapes and colours: The following default shapes are accessible via the Tools menu: rectangle, line, arrow, ellipse, polygon, star, images. Hold Shift to maintain proper proportions. Pen Instrument: To express one's creativity or to depict a complicated form, the pen tool may be employed (press P). Colours: Choose the shape's colour from the "Fill" section of the menu on the right.

Layers and groups: They emerge automatically upon the creation of an element. They are accessible via the menu on the left. Nesting frames within other frames is also possible. Also, the layers may be grouped using cmd+G which helps in keeping the files better organised.

Pages and images: Pages are accessible via the menu on the left and may be used for various project components, such as wireframes and UIs. Users may effortlessly import photographs into the canvas by dragging and dropping them into it (Figure 10.13). By clicking the picture icon in the Inspect panel, Figma enables modifications such as contrast, saturation, and exposure.

Typography and styles: Commence writing by pressing T or searching the Tools menu for the icon. Although all Google fonts are pre-installed in Figma, users may install Figma Font Helper to utilise your own desktop fonts if you choose. In the Inspect panel, users may modify text attributes including size, colour, letter spacing, and more. A very useful Figma feature, styles enables the saving of reusable characteristics like as grids, colours, text, and shadows. Working through the frames while depending only on

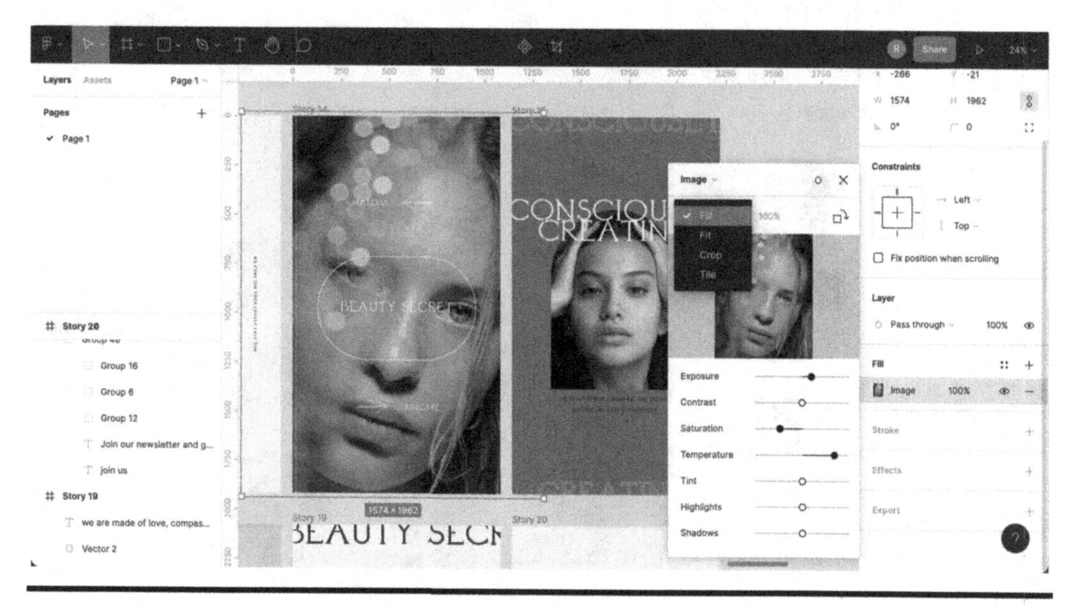

Figure 10.13 Layers and groups [26].

these definitions can save users an enormous amount of time. To name and store a colour style, select the desired hue, navigate to the Inspect panel under Fill, and then click the button denoted by four dots labelled "Styles." Subsequently, press the plus sign (Figure 10.14).

Components and constraints: Components are aspects that aid in the establishment of project uniformity. Master components and instances both exist. Property values are established by the master component; modifications to this component will have an impact on every instance of it. Select a collection of layers and either right-click to create component or press opt+cmd+K to generate a component. Constraints dictate the behaviour of elements during frame resizing. The default configuration for the restrictions is top and left. Users may modify these in the Constraints section of the Inspect panel by choosing the items in your frame. Both the x (horizontal) and y (vertical) axes are applicable (Figure 10.15).

Prototyping: Creating a prototype is an extremely important process for determining how customers will interact with a proposed product. Prototypes may be seen and connections, interactions, and animations can be generated. Here are some definitions at a minimum: interconnections are established between frames and their constituent pieces. To initiate the connecting process, ensure that the designer chooses "Prototype" from the Inspect panel. Click on an element or frame, a circle will appear. In order to redirect the connection to a new frame, click and hold.

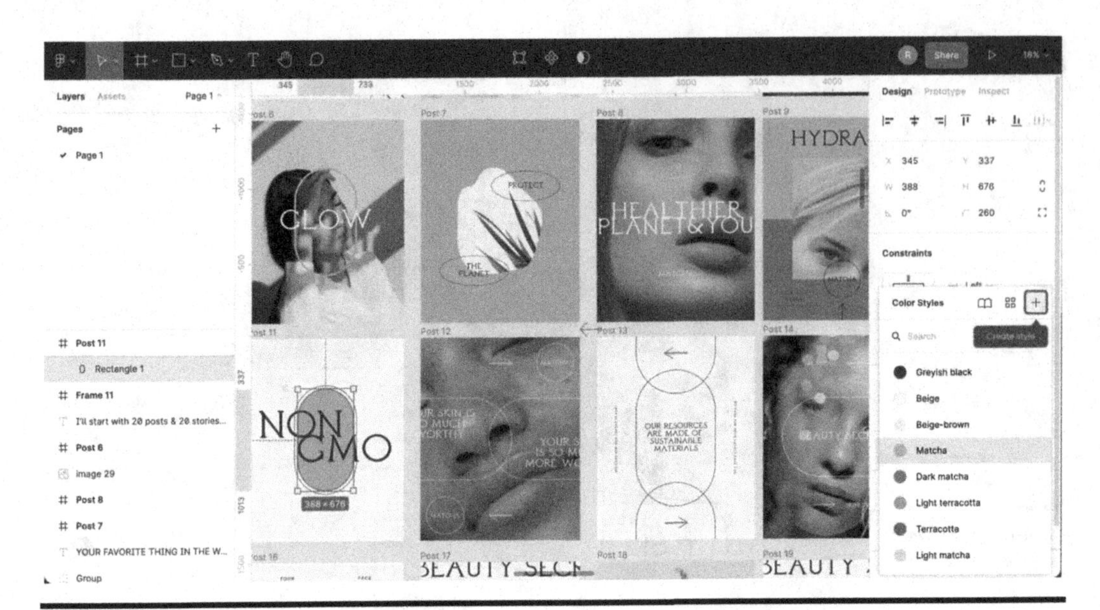

Figure 10.14 Figures and styles [26].

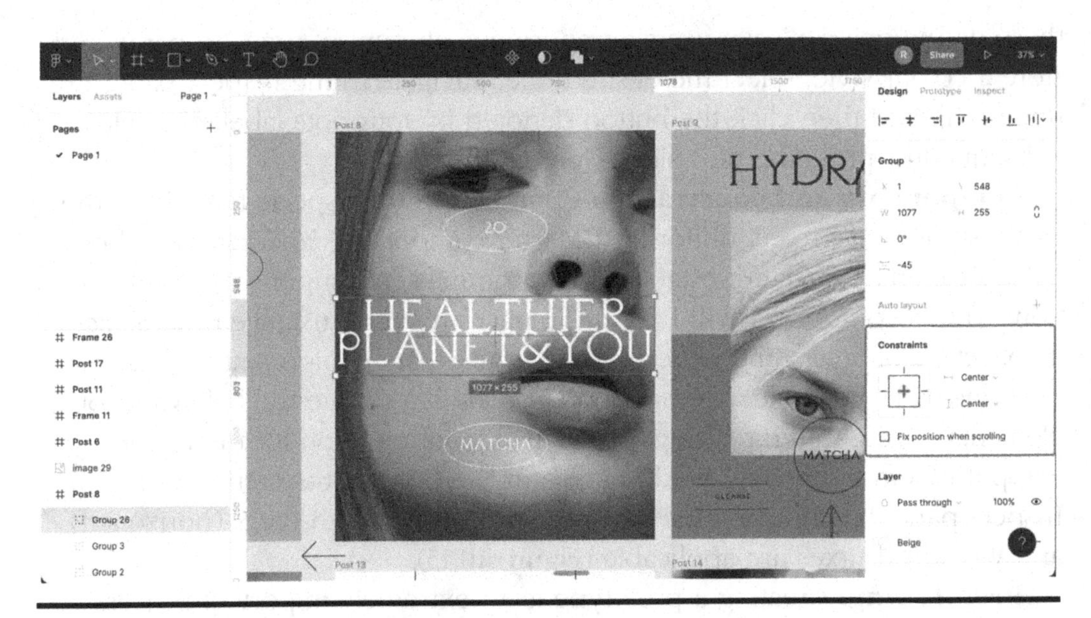

Figure 10.15 Components and constraints [26].

Animations and interactions: The interaction details will be displayed once the connection has been established. Select from an extensive library of animations and interactions. **Prototypes:** To begin interacting with the prototype, use the forward arrow button located in the upper-right corner of the display.

References

1. Lee S, Geum Y. How to determine a minimum viable product in app-based lean start-ups: Kano-based approach. *Total Qual Manag Bus Excell* 2021;32:1751–67. https://doi.org/10.1080/14783363.2020.1770588.
2. How to build an MVP in healthcare: process, cost, and benefits n.d. https://radixweb.com/blog/build-an-mvp-in-healthcare (accessed December 15, 2023).
3. MVP In healthcare: five steps to avoid chaos n.d. https://www.forbes.com/sites/forbestechcouncil/2023/04/11/mvp-in-healthcare-five-steps-to-avoid-chaos/?sh=60817678c3fc (accessed December 15, 2023).
4. Minimum viable product - what is a MVP and why is it important? n.d. https://www.productplan.com/glossary/minimum-viable-product/ (accessed December 15, 2023).
5. Amante AD, Ronquillo TA. Technopreneurship as an outcomes-based education tool applied in some engineering and computing science programme. *Australas J Eng Educ* 2017;22:32–8. https://doi.org/10.1080/22054952.2017.1348186.
6. How to build an MVP for healthcare product: 4 Key steps n.d. https://demigos.com/blog-post/how-to-build-an-mvp-for-healthcare-product/ (accessed December 15, 2023).
7. Dennehy D, Kasraian L, O'Raghallaigh P, Conboy K, Sammon D, Lynch P. A lean start-up approach for developing minimum viable products in an established company. *J Decis Syst* 2019;28:224–32. https://doi.org/10.1080/12460125.2019.1642081.
8. Cook DA, Bikkani A, Poterucha Carter MJ. Evaluating education innovations rapidly with build-measure-learn: Applying lean startup to health professions education. *Med Teach* 2023;45:167–78. https://doi.org/10.1080/0142159X.2022.2118038.
9. Healthcare apps available Google Play 2022 | Statista n.d. https://www.statista.com/statistics/779919/health-apps-available-google-play-worldwide/ (accessed December 15, 2023).
10. How to create a healthcare mvp n.d. https://zenbit.tech/blog/how-to-create-a-healthcare-mvp/ (accessed December 15, 2023).
11. What is an MVP & how to apply them in healthcare n.d. https://kms-healthcare.com/what-is-an-mvp-healthcare-industry/ (accessed December 15, 2023).
12. Minimum viable product: healthcare innovation unleashed n.d. https://www.moonshot.partners/blog/minimum-viable-product-healthcare-innovation-unleashed (accessed December 15, 2023).
13. MVP for a healthcare startup and how to create one l declaration n.d. https://decklaration.com/how-to-create-mvp-for-a-healthcare-startup/ (accessed December 15, 2023).
14. Lillrank P. The logics of healthcare: the professional's guide to health systems science. *Logics Healthc Prof Guid to Heal Syst Sci* 2018:1–288. https://doi.org/10.1201/B22039/LOGICS-HEALTHCARE-PAUL-LILLRANK.

15. Sparviero S. The case for a socially oriented business model canvas: the social enterprise model canvas. *J Soc Entrep* 2019;10:232–51. https://doi.org/10.1080/19420676.2018.1541011.
16. How to: business model canvas explained | by Sheda | Sheda | Medium n.d. https://medium.com/seed-digital/how-to-business-model-canvas-explained-ad3676b6fe4a (accessed December 15, 2023).
17. Business model canvas – download the official template n.d. https://www.strategyzer.com/library/the-business-model-canvas (accessed December 15, 2023).
18. Rytkönen E, Nenonen S. The business model canvas in university campus management. *Intell Build Int* 2014;6:138–54. https://doi.org/10.1080/17508975.2013.807768.
19. Beck Dallaghan GL, Lomis K, Crow S, Coplit L. Bridging educational innovation and financial offices: using the business model canvas modified for medical educators to communicate need. *J Commun Healthc* 2022;15:131–6. https://doi.org/10.1080/17538068.2021.1993691.
20. Prado AM, Pearson AA, Bertelsen NS. Management training in global health education: a health innovation fellowship training program to bring healthcare to low-income communities in Central America. *Glob Health Action* 2018;11. https://doi.org/10.1080/16549716.2017.1408359.
21. Hospital business model canvas [classic] | Creately n.d. https://creately.com/diagram/example/ihhwbin81/hospital-business-model-canvas-classic (accessed December 15, 2023).
22. Roth RE, Hart D, Mead R, Quinn C. Wireframing for interactive & web-based geographic visualization: designing the NOAA lake level viewer. *Cartogr Geogr Inf Sci* 2017;44:338–57. https://doi.org/10.1080/15230406.2016.1171166.
23. Kyratzi S, Azariadis P. Geometric definition of the hidden part of a line drawing in a sketch-to-solid methodology. *Comput Aided Des Appl* 2015;12:355–65. https://doi.org/10.1080/16864360.2014.981466.
24. Domingos Alves D, Souza Matos E de, Chavez C von FG . Interaction design in distributed software development: a systematic mapping study. *Behav Inf Technol* 2023. https://doi.org/10.1080/0144929X.2023.2286534.
25. Figma: the collaborative interface design tool n.d. https://www.figma.com/ (accessed December 15, 2023).
26. Figma basics: a guide for beginners - creative market blog n.d. https://creativemarket.com/blog/figma-guide (accessed December 15, 2023).
27. Wireframe: What is a Wireframe. https://www.productplan.com/glossary/wireframe. (accessed December 12, 2023).

Chapter 11

Manufacturing Techniques and Material Selection for Healthcare Product Prototyping

11.1 Manufacturing Techniques

The manufacturing of medical devices encompasses every stage of the fabrication process, from process design to scale-up and continuous process improvements. In addition, sterilisation and shipment of the gadget are also included [1, 2]. In addition to striving for greater speed and efficiency throughout the production process, medical device manufacturers also aspire to be accountable corporate citizens. Therefore, manufacturing necessitates ongoing awareness of sustainable materials, renewable resources, energy-efficient equipment, and waste reduction strategies [3]. Potential remedies for these challenges may manifest as enhanced operational procedures, technical progressions in machinery or equipment components, or materials that are more secure and dependable [2]. The packing process is governed by identical principles.

Numerous medical device manufacturers outsource the manufacturing of components or full devices to contract manufacturers because they excel in the ideation, concept, and prototype stages of product development. This is true for both established original equipment manufacturers (OEMs) and mid-sized and startup businesses [2]. Additionally, contract manufacturers differ in scale and level of proficiency; many establishments are compact, specialised

DOI: 10.4324/9781003475309-15

manufacturing facilities that focus on certain materials or components, whilst others are expansive cleanroom complexes outfitted for mass production.

The manufacture of medical devices is an essential sector that exerts a substantial influence on our healthcare system. The industry produces and distributes a broad selection of medical devices that find utility in several contexts, including pacemakers, medical stylets, X-ray and MRI scanners, and more [4]. The manufacturing of medical devices, as an industry, may involve any or all of the production processes, including packaging, customised shipping, and design and fabrication. As part of a larger medical manufacturing sector, medical gadgets, tools, and equipment utilised in healthcare are produced. This sector encompasses two overarching domains. The initial sector pertains to the manufacture of medical devices. This encompasses every facet of the fabrication, design, and manufacturing procedures associated with the creation of a medical device. The second domain pertains to the advancement of techniques and technology utilised in the medical manufacturing sector [4]. Medical device production continues to increase as a result of technological breakthroughs and innovation in the fields of material sciences, which are constantly expanding.

The Food and Drug Administration, or FDA for short, regulates the production of medical devices in the United States in order to guarantee their safety and effectiveness [4]. In the context of medical devices, it is imperative that manufacturers adhere to and satisfy FDA regulatory mandates and device-specific standards throughout the whole process of product design, manufacturing, and marketing. A medical device is defined by the FDA as any apparatus, device, instrument, equipment, or delivery system that is utilised or employed for a medical purpose or application, with the explicit objective of eliciting non-chemical effects on the body, while implantable medical devices are incapable of undergoing absorption or metabolism by the body.

Therefore, medical gadgets are vital to public health. In addition to testing, monitoring, analysing, and diagnosing illnesses and other biological human problems, they are employed for treatment and prevention. Medical device development and manufacturing firms play a crucial and indispensable role in the biological sciences, health and healthcare, and all medical-related sectors and disciplines. The production of customised medical devices caters to an extensive array of biotechnology sectors. In addition to saving time and resources for the primary manufacturer, these manufacturers typically possess superior staff, production capabilities, materials, and equipment required to make a medical device component or product.

Utilised in medical devices and applications, spring components and wire forms constitute one such field of specialised medical manufacturing. Surgical instruments and medical devices that operate via mechanisms will have a spring component. Medical devices that rely on springs for operation include apparatus for medical testing and monitoring, MRI and X-ray machinery, dental X-rays featuring articulated arms, various types of booms, table and bed lifts, wheelchairs, fluid control and delivery devices, anaesthetic infusion devices, sterilisation valves, various implants, shunt valves, and even mechanisms for closing medical equipment drawers [4]. A multitude of medical devices, including spinal implant rods, medical slips, and several sorts of medical stylets, are manufactured with wire forms. Additionally, the appropriate material qualities and component shape must be addressed. For intricate part geometries, additive manufacturing and injection moulding are often more favourable, although the material characteristics differ among the three groups. A summary of more information on prevalent methods utilised in the production of medical equipment, including additive manufacturing, computer numerical control machining, and injection moulding, is provided below.

11.1.1 Computer Numerical Control Machining

Computer numerical control (CNC) machining is a subtractive manufacturing process in which components are produced by utilising a range of cutting tools to remove material from a solid block (called the blank or the workpiece) [5]. This manufacturing process is inherently distinct from formative (injection moulding) and additive (3D printing) technologies. The ramifications of the material removal techniques on the advantages, constraints, and design limits of CNC are substantial. Figure 11.1 shows a typical subtractive manufacturing technique wherein the revolving tool cuts off the metal part as per the required design.

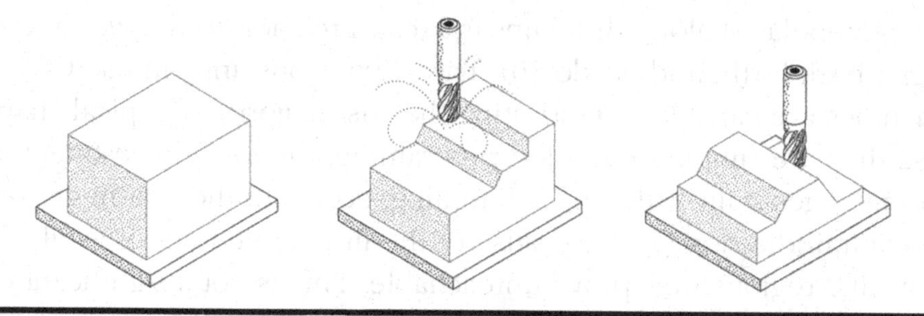

Figure 11.1 Schematic of a typical subtractive manufacturing technique [6].

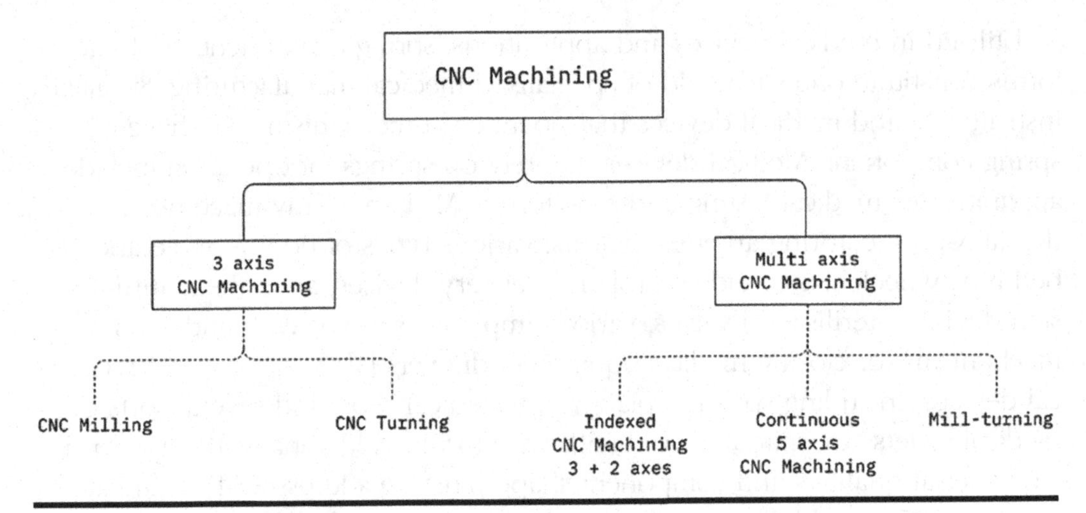

Figure 11.2 Different types of CNC machines [6].

CNC machining is a digital manufacturing process that generates flawlessly precise components with exceptional physical characteristics by utilising a computer-aided design (CAD) file as input. CNC, by virtue of its extensive automation, offers reasonable pricing for both one-of-a-kind bespoke components and medium-volume products [7, 8]. The fundamental CNC procedure consists of three phases. The engineer begins by creating the component's CAD model. After converting the CAD file to a CNC programme (G-code), the machinist configures the machine. In conclusion, the CNC system performs material removal and component fabrication with little oversight across all machining processes. The CNC machines are typically of two types, three-axis machines and five-axis machines. Figure 11.2 shows different types of CNC machines wherein three axis means that the toolhead could perform manoeuvres in three different axes whereas a five-axis machine could perform machining in five different axes.

CNC turning (Figure 11.3) and milling machines are two types of CNC systems that utilise three axes. The cutting tool of these "basic" machines may be manipulated along three linear axes in relation to the workpiece (left-right, back-forth, and up-down). Three-axis constantly-in-use CNC milling machines are capable of fabricating the vast majority of typical shapes. Because they are simple to use and programme, the initial investment in machining is generally little. In CNC milling, access to the tool might be a design contraindication [7, 8]. Because of the limited number of available axes, specific regions may prove unreachable. This is not a significant concern if the workpiece only requires a single turn; but, labour and machining expenses quickly grow when many rotations are required. Figure 11.4 shows

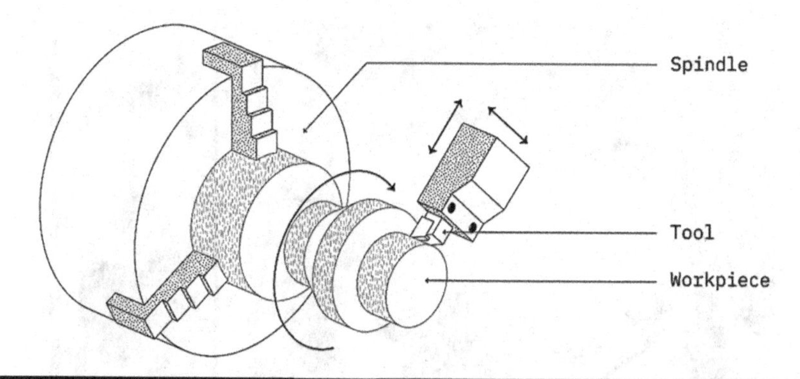

Figure 11.3 Schematic of a CNC turning machine [6].

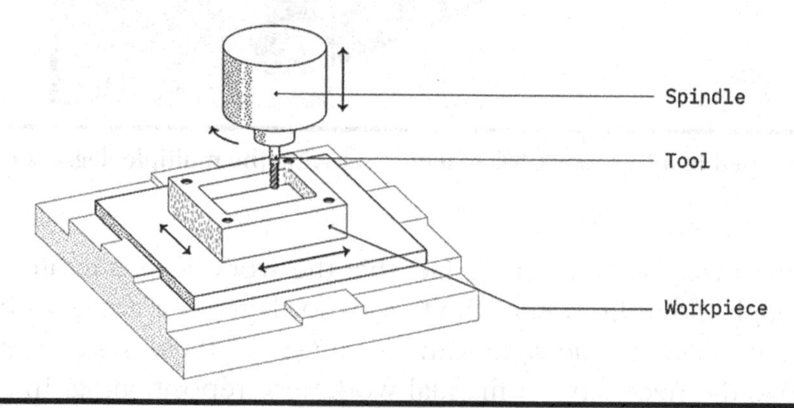

Figure 11.4 Working of a three-axis CNC milling machine [6].

the working of a three-axis CNC machine where the spindle holds the tool and performs the machining on the workpiece.

Compared to three-axis CNC, a five-axis CNC has three varieties: mill-turning centres with live tooling, continuous five-axis CNC milling, and five-axis indexed CNC milling. In essence, these systems are lathes or milling machines with extra degrees of freedom. Five-axis CNC milling centres, for instance, permit rotation along the toolhead, machine bed, or both, in addition to the three linear axes of motion. The enhanced functionalities of these computers are accompanied with a heightened price. In addition to specialised equipment, they need operators possessing specialist expertise. However, in the case of metal components that are extremely intricate or require topological optimisation, 3D printing is often a more viable alternative.

The cutting tool can only rotate along three linear axes when machining. While performing operations, the toolhead and bed may spin, providing an alternative vantage point from which to observe the workpiece. Indexed five-axis CNC milling systems are sometimes referred to as 3 + 2

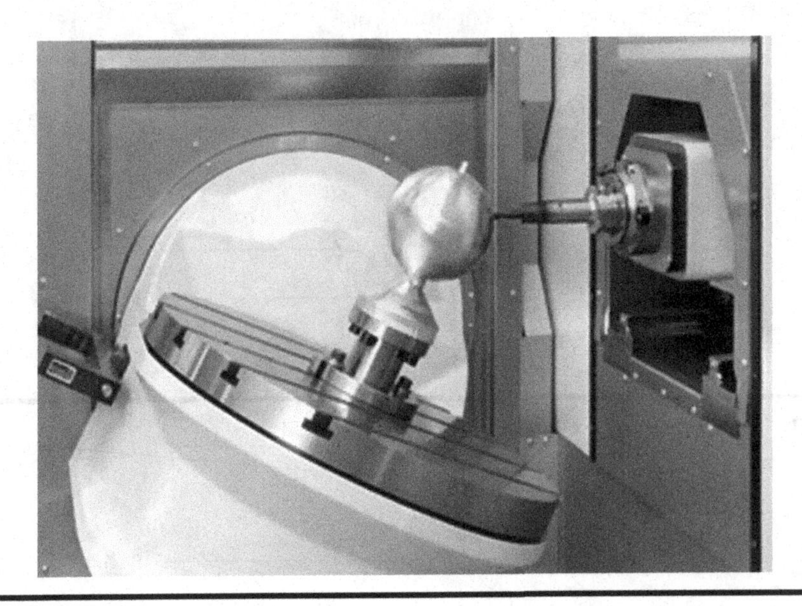

Figure 11.5 Bed of a five-axis CNC machine showing the multiple degrees of freedom [6].

CNC milling machines due to the fact that the workpiece is rotated exclusively through the utilisation of the two extra degrees of freedom between machining processes. One significant advantage of these systems is that they remove the necessity for manual workpiece repositioning. In this manner, components featuring more intricate geometries may be produced with more velocity and precision compared to a three-axis CNC machine. Figure 11.5 shows the bed of a five-axis CNC machine showing the degree of freedom.

The machine design of continuous five-axis CNC milling systems is comparable to that of indexed five-axis CNC milling machines. However, they enable simultaneous motion along all five axes during the entirety of the machining process. By employing this method, components featuring intricate, "organic" geometries may be created to an extent that is unattainable by the use of alternative technologies. The expense of these sophisticated functionalities is unsurprising, given that they need both costly gear and extensively educated machinists.

Figure 11.6 shows a typical CNC turning machine. A spindle, on which the workpiece is mounted, has the capability to either rotate rapidly (as in a lathe) or precisely angle it (like a five-axis CNC mill). Mill-turning CNC centres consist mostly of CNC lathe machines that are complemented by CNC milling tools. Swiss-style lathes, an alternative to mill-turning centres, are

Figure 11.6 A typical CNC turning machine [6].

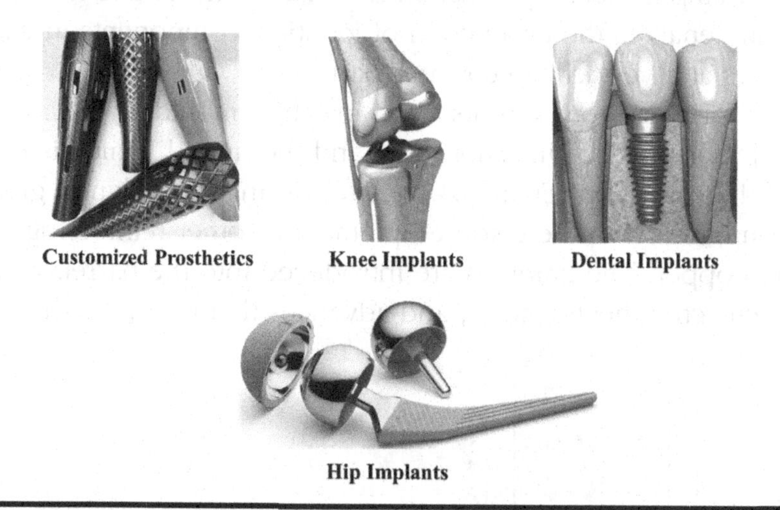

Figure 11.7 Application of CNC machining in the healthcare industry.

characterised by a generally greater precession. Mill-turning systems use the geometric freedom of CNC milling and the high productivity of CNC turning. Consider camshafts and centrifugal impellers; they produce components with "loose" rotational symmetry at a much-reduced cost compared to other five-axis CNC machining technologies. CNC machining has numerous applications in the medical industry such as customised prosthetics, knee, hip, and dental implants, etc. (Figure 11.7).

11.1.2 Injection Moulding

Injection moulding is a manufacturing process utilised to mass-produce plastic components that are similar and possess favourable tolerances. Melted polymer grains are pumped under pressure into a mould for injection moulding, where the liquid plastic hardens after cooling. Injection moulding utilises thermoplastic polymers that are capable of acquiring additional colours or fillings. Injection moulding is extremely popular because of the extremely cheap unit cost that results from producing in large quantities. Injection moulding provides excellent design freedom and great reproducibility. Economic factors frequently dictate the limitations of injection moulding, given the substantial initial investment required for the mould. Furthermore, there is a gradual progression from design to manufacturing (at least four weeks).

Injection moulding is currently a prevalent technique employed in the production of technical and consumer goods. Almost every plastic object in the vicinity was produced by injection moulding. This is due to the fact that the technique enables the production of identical components in extremely high numbers at a very cheap cost per part.

The three primary components of an injection moulding machine are the clamping/ejector unit, the injection unit, and the mould, which is the central element of the operation (Figure 11.8). After drying the polymer granules, they are combined with the coloured pigment or other reinforcing additives in the hopper. The granules are introduced into the barrel, where a variable pitch screw heats, mixes, and advances them towards the mould

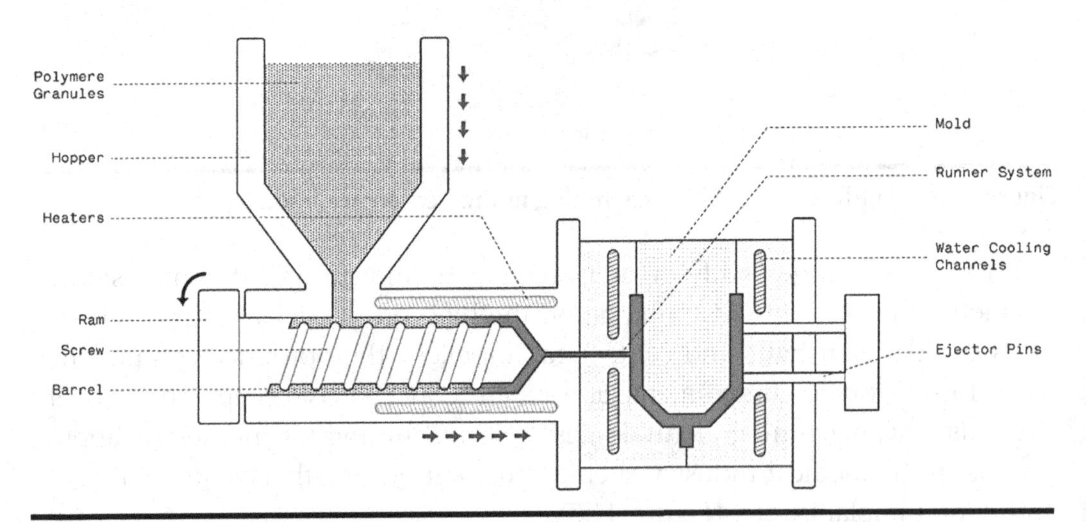

Figure 11.8 Injection moulding [9].

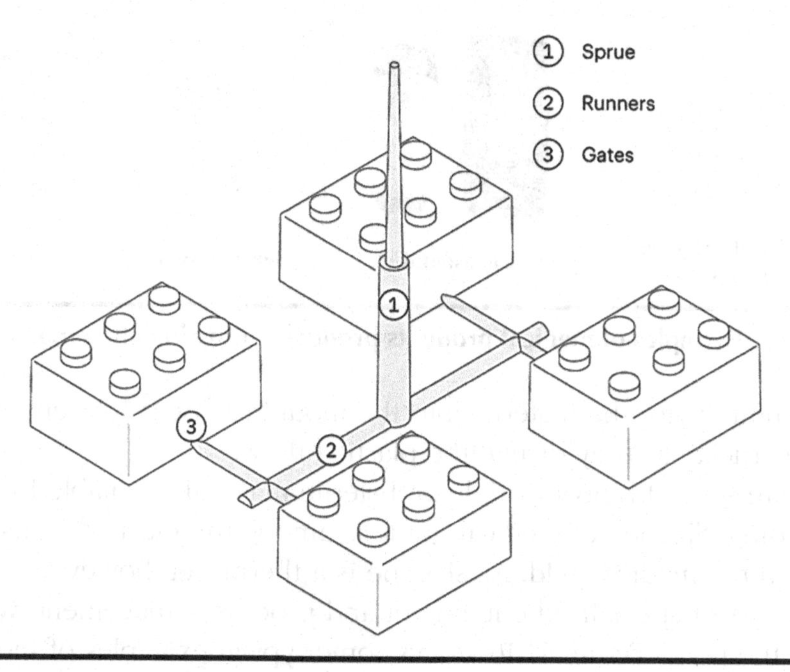

Figure 11.9 Internal designs through injection moulding [9].

concurrently. The optimisation of the screw and barrel design facilitates the generation of ideal pressure levels, hence enabling the material to be melted. The runner system then forces the molten plastic into the mould through the ram, where it completely fills the cavity. Resolidifying as the substance descends, it assumes the contour of the mould. Ultimately, the mould is ajar, and the ejector pins expel the now-solid component. The procedure is then repeated when the mould is sealed. The cycle may be replicated in a relatively short amount of time: between 30 and 90 seconds, depending on the size of the component.

If there are internal designs, those are manufactured using a runner system (Figure 11.9). The route via which the molten plastic is guided into the mould cavity called the runner system. It regulates the pressure and flow of the liquid plastic that is injected into the cavity; it is eliminated following ejection (it snaps off). As the molten plastic enters the mould, it initially runs through the sprue, which serves as the primary conduit.

Connecting the sprue to the gates, the runner distributes the molten plastic along the face where the two parts of the mould meet. One or more runners may be present, instructing the content to go to one or more distinct sections. Following ejection, the runner system is severed from the component. Only this type of material waste occurs during injection moulding, and between 15 and 30% of it is recyclable or reusable. The gate serves as the

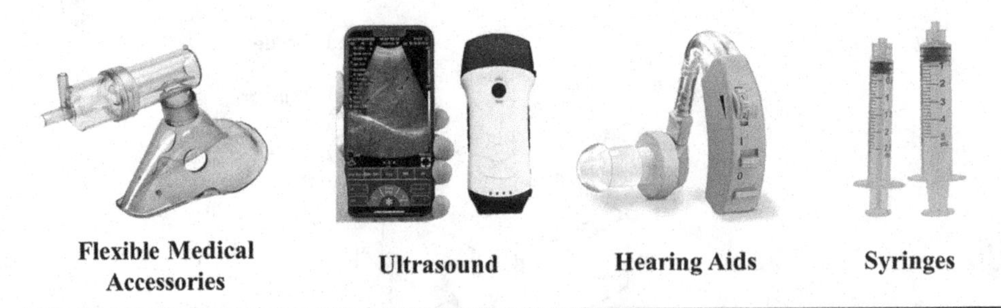

Figure 11.10 Examples of medical products produced using injection moulding.

point of entrance for the material into the mould cavity. Its placement and shape are crucial, as they dictate the plastic's flow.

There are several biocompatible and sterile materials available for injection moulding. Silicone of medical grade is among the most often used materials in the medical field. As silicone is a thermoset, however, it necessitates the use of specialised equipment and process management, which drives up the price. Figure 11.10 shows some typical examples of medical products produced using injection moulding.

11.1.3 3D Printing

11.1.3.1 Fused Deposition Modelling

Fused deposition modelling (FDM) 3D printing, sometimes referred to as fused filament fabrication (FFF), operates as an additive manufacturing (AM) methodology with material extrusion as its domain [10]. Layer by layer, FDM constructs components by depositing molten material selectively along a specified route [11]. Filaments of thermoplastic polymers are utilised in the fabrication of the final physical items [12]. 3D printers construct parts by layering molten filament material over a build platform until the component is complete [13]. FDM converts digital design files into physical dimensions via the machine itself when the files are submitted. Polymers such as ABS, PLA, PETG, and PEI are utilised in FDM materials. Out of these, PLA is often used in medical devices as it is easy to print [14–17]. These substances are fed into the machine in the form of threads via a heated nozzle.

Before using an FDM machine, a spool of this thermoplastic filament must be loaded into the printer. Once the nozzle reaches the proper temperature, the filament is fed via an extrusion head and nozzle by the printer [18]. The three-axis system to which this extrusion head is mounted enables it to rotate along the X, Y, and Z axes (Figure 11.11). Layer by layer, the

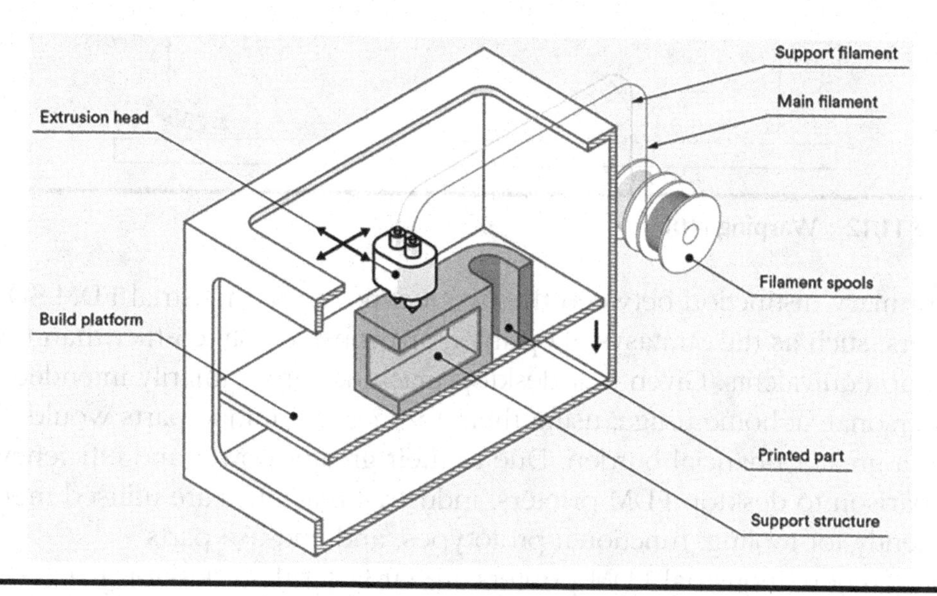

Figure 11.11 Basic structure of a cartesian-based 3D printer [10].

printer deposits molten material extruded in thin strands along a design-determined route. The deposited substance solidifies and cools [19]. Fan attachment to the extrusion head may be employed in certain circumstances to expedite the cooling process. Similar to how numerous passes are necessary to colour in a shape with a marker, filling in an area requires such action. The build platform falls when the printer completes a layer, at which point the machine begins operation on the subsequent layer. In some machine configurations, the extrusion head ascends. This procedure is iterated until the component is complete.

The majority of FDM systems provide the modification of several process parameters. The nozzle and build platform temperatures, build speed, layer height, and cooling fan speed are some of these variables. Typically, designers are generally not concerned with these modifications, since an AM operator has already taken care of them. It is crucial to take into account two significant factors, namely the construction size and layer height. Industrial machines may achieve build sizes of 1,000 × 1,000 × 1,000 mm, whereas the standard dimensions for consumer 3D printers are 200 × 200 × 200 mm. The customary range for FDM layer height is from 50 to 400 microns. While printing shorter layers yields smoother components and captures curved geometries more precisely, printing taller layers expedites the process and reduces costs.

There are typically two primary classifications for FDM printers: industrial (professional) and prototype (desktop) equipment. While both printer classes provide unique uses and benefits, their production scales constitute

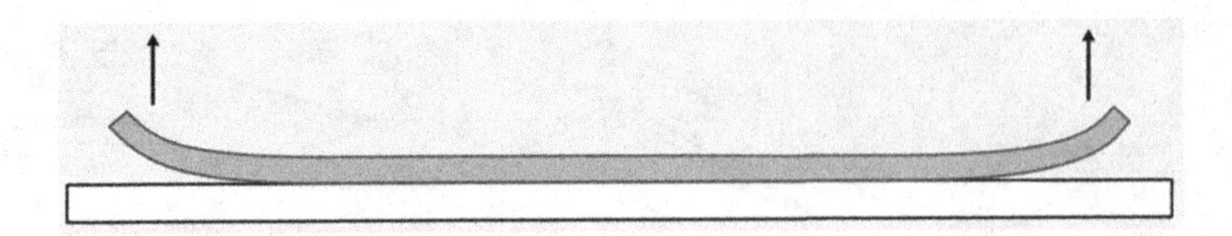

Figure 11.12 Warping [10].

the primary distinction between the two technologies. Industrial FDM 3D printers, such as the Stratasys 3D printer, are considerably costlier than their desktop equivalents. Given that desktop machines are primarily intended for personal, at-home usage, using them to produce unique parts would incur a greater financial burden. Due to their greater power and efficiency in comparison to desktop FDM printers, industrial machines are utilised more frequently for tooling, functional prototypes, and end-use parts.

In addition, industrial FDM printers are far faster than desktop machines in producing bigger orders. They are intended for dependability and reproducibility and need minimum human interaction to manufacture identical parts again. Desktop FDM printers lack significantly more durability. It is necessary to do routine calibration and user maintenance on desktop computers.

There are some characteristics of FDM printers that needs to be understood by the designers and entrepreneurs:

Warping: Warping is among the most prevalent FDM flaws (Figure 11.12). As extruded material solidifies and cools, its dimensions diminish. As distinct regions of the printed component have differential cooling rates, their respective dimensions likewise undergo differential rate changes [20]. Internal tensions accumulate as a result of differential cooling; these strains pull the underlying layer upward, causing it to distort. Multiple methods exist for preventing warping. Keeping a tight eye on the temperature of the FDM system, particularly the build platform and chamber, is one approach. Additionally, warping can be reduced by increasing the adhesion between the component and the build platform.

Layer adhesion: In FDM, the adhesion between deposited layers of a component is vital. The molten thermoplastic that is extruded through the nozzle of an FDM machine exerts pressure on the layer that was previously printed (Figure 11.13). By re-melting this layer at high temperature and pressure, it becomes possible for it to form a bond with the preceding layer. The previously printed layer undergoes a deformation into an oval shape as a result of the molten material's pressure. This implies that irrespective of the layer height employed, FDM components invariably have an undulating

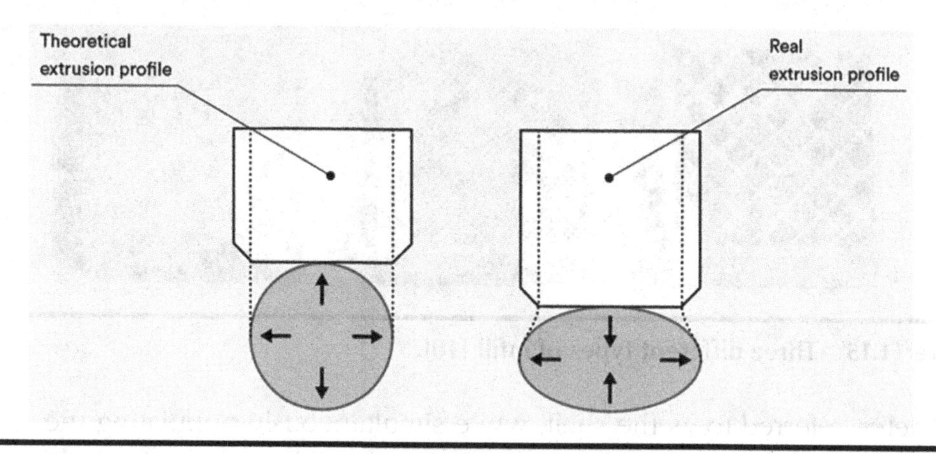

Figure 11.13 Layer adhesion [10].

Figure 11.14 Support structures [10].

surface, and even minute characteristics like threads or holes may necessitate further post-processing.

Support structures: It is not feasible for FDM printers to discharge molten thermoplastic into thin air. Support structures, which are frequently printed in the same material as the components, are necessary for specific part geometries. Eliminating support structure materials can sometimes provide challenges; however, designing components to reduce the need for support structures is frequently a more straightforward approach. There are liquid-soluble support materials available, however they are often utilised in conjunction with more advanced FDM 3D printers. Consider that the use of dissolvable supports will result in a higher total cost for printing. Figure 11.14 shows the 3D printed part having support structures.

Infill and shell thickness: In order to conserve resources and decrease printing time, FDM printers often do not manufacture solid pieces. On the contrary, the machine proceeds in many passes to trace the outside

Figure 11.15 Three different types of infill [10].

perimeter, referred to as the shell, while simultaneously populating the inside, dubbed the infill, with a low-density internal structure. Shell thickness and infill weight have a substantial impact on the robustness of FDM-printed components. The default infill density and shell thickness of the majority of desktop FDM printers are 20% and 1 mm, respectively, which provides an adequate balance of strength and speed for rapid printing. Figure 11.15 shows three different types of infills. The left most had less infill compared to others, it would take less time to print. Also, it consumed less material and was eventually lighter compared to the other two. Designers could choose the infill percentage based on the required product strength.

Post-processing: Removal of supports, metal plating, priming, sanding, cold welding, vapour smoothing, epoxy coating, and priming are a few of the post-processing techniques that may be used to get a highly polished FDM 3D-printed component finish.

11.1.3.2 Stereolithography (SLA)

Stereolithography (SLA) is the most prevalent resin printing technology and a frequently employed 3D printing method. The technology has gained recognition in the additive industry because of its capability of producing precise, waterproof, and isotropic prototypes, in addition to manufacturing components with exceptional surface smoothness and intricate details. Although SLA has several benefits, it might be difficult to determine whether it will produce optimal results for your particular components. SLA is among the most used processes for vat photopolymerisation. By selectively curing a polymer resin layer by layer with an ultraviolet (UV) laser beam, things are produced. SLA utilises liquid photosensitive thermoset polymers as its constituent ingredients [21]. SLA remains the most economically viable 3D printing process when extremely precise components or those requiring a clean

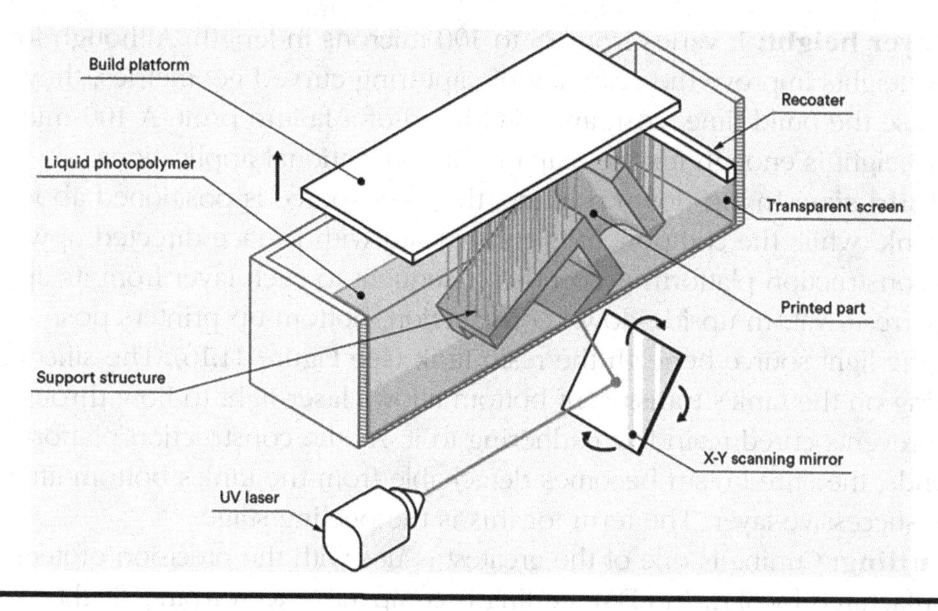

Figure 11.16 Working mechanism of a SLA based 3D printer [22].

surface finish are required. A designer achieves optimal outcomes by capitalising on the advantages and constraints inherent in the production process. To consider these aspects, it is important for a designer or entrepreneur to fully understand the working of the SLA printing.

SLA 3D printing operates by initially putting the build platform within the liquid photopolymer tank, one layer height above the liquid's surface (Figure 11.16). The subsequent layer is produced by a UV laser curing and hardening the photopolymer resin in a selective manner. The solidification phase of photopolymerisation involves the activation of the monomer carbon chains comprising the liquid resin by the UV laser. This results in the chains solidifying and forming strong, unbreakable links among themselves. The laser beam is directed along a pre-established trajectory by means of a collection of mirrors known as galvos. By scanning the whole cross-sectional area of the model, the resulting component is completely solid. The printed component remains in an incompletely cured condition. Additional post-processing under UV radiation is necessary to achieve exceptionally excellent mechanical and thermal characteristics.

There are several print parameters that need to be considered before getting started with the process. The majority of print settings in SLA systems are manufacturer-fixed and cannot be altered. Only the layer height and component orientation are required as inputs (the latter determines support location).

Layer height: It varies from 25 to 100 microns in length. Although lower layer heights improve the accuracy of capturing curved geometries, they also increase the build time, cost, and likelihood of a failure print. A 100-micron layer height is enough for the majority of conventional applications.

Build size: In top-down printing, the laser source is positioned above the tank, while the component is constructed with its face directed upwards. The construction platform descends subsequent to each layer from its apex in the resin vat. In upside-down construction, bottom-up printers position the light source beneath the resin tank (see Figure 11.16). The silicone coating on the tank's translucent bottom allows laser light to flow through but prevents cured resin from adhering to it. As the construction platform ascends, the cured resin becomes detachable from the tank's bottom after each successive layer. The term for this is the peeling stage.

Curling: Curling is one of the greatest issues with the precision of items manufactured by SLA. In FDM, curling is comparable to warping. Slight shrinkage of the resin occurs during the curing process when it is exposed to the light source of the printer. Significant internal tensions form between the newly formed layer and the previously solidified material when the shrinkage is substantial; this causes the component to curl. The presence of support is crucial in order to stabilise vulnerable regions of a print to the build plate and reduce the probability of curling. Additionally crucial are part orientation and the restriction of thick, flat layers. Additionally, excessive curing (such as exposing the component to direct sunlight after printing) may result in curling. It is most effective to avoid curling by keeping this in mind throughout the design process. Whenever feasible, square or flat areas should be avoided, or a structure should be added to prevent the area from curling.

Layer adhesion: Components produced with SLA have isotropic mechanical characteristics. This is because the liquid resin cannot be completely cured with a single UV laser pass. Significantly enhanced fusion of previously cemented layers is facilitated by subsequent laser passes. Indeed, curing persists even subsequent to the printing process reaching its conclusion. In order to attain optimal mechanical qualities, SLA components must be post-cured in a cure box containing powerful UV radiation (and sometimes at elevated temperatures). This significantly increases the SLA component's temperature resistance and hardness but makes it more brittle.

Medical device applications include low-cost prototypes, surgical and dental instruments and guides, and some biomedical implants, such as hearing aids, when high material strength is not required. Some of the examples are presented in Figure 11.17.

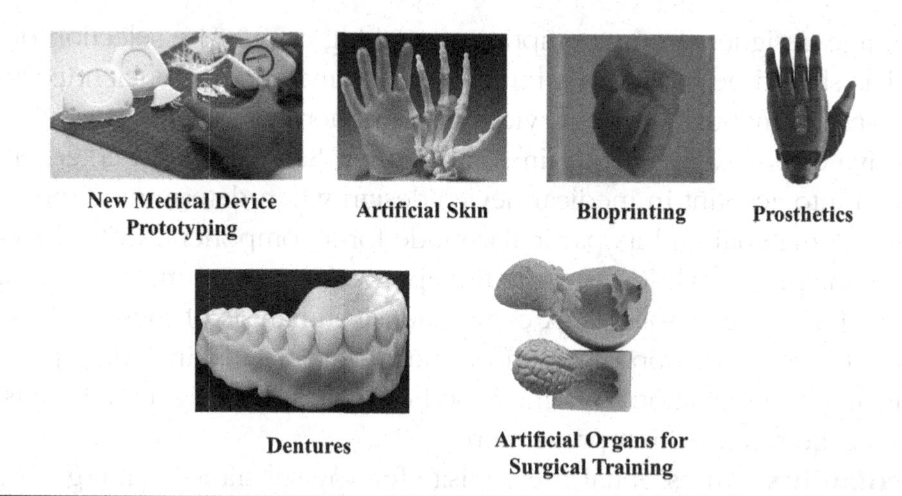

Figure 11.17 Some applications of SLA printing in the healthcare industry [23].

11.2 Material Selection in Healthcare Product Development

A vital element in the development of a medical device is the selection of the proper material for each component. This requires knowledge of concerns ranging from budgetary limits and supply chain logistics to physical performance and manufacturing limitations. As a result of innovative medical technology, patient care, therapies, and diagnostics are all enhanced. However, these state-of-the-art advancements cannot be realised without the proper substance.

The selection of materials for incorporation into novel medical devices can have significant consequences at several stages of the product lifecycle, including design, testing, regulatory approvals, manufacture, assembly, market acceptance, and eventual disposal. Metals, polymers, composites, and elastomers must be consistently and easily accessible throughout the supply chain. The selected suppliers ought to ideally provide superior performance and safety while adhering to the intended price range. Additionally, the materials ought to provide beautiful, pleasant, and ergonomic solutions, encourage regulatory compliance, facilitate production, and boost creative flexibility, all while appealing to providers and patients. Material specifications that check the greatest number of these boxes can have a significant impact on time to market and commercial success.

Nonetheless, in light of the inexorably vast array of material options and the design elements that are unavoidable, the choosing process can become exceedingly intricate and time intensive. This difficulty is exacerbated by device manufacturers without in-depth knowledge of materials, and it may discourage designers from attempting something novel. The selection of materials should be based on a number of parameters. A proficient partner in design and manufacturing services may frequently assist in preventing expensive errors, so save both time and money. Several critical criteria must be taken into account in medical device design when determining the suitability of a material and its particular grade for a component. Critical factors to contemplate while assessing the appropriateness of materials for a medical device are as follows. A comprehensive analysis of these factors necessitates consideration of significant healthcare trends, including population ageing, the transition to home-based care, and the heightened focus on hospital-acquired infection prevention.

Availability: An essential prerequisite for any substance is a regular and dependable supply. Insufficient availability of quantities to meet present and future demands may have an adverse effect on the time to market and sales potential of a novel device. Additionally crucial is the proximity of supply production to the device manufacturing location, which should be optimised to minimise transportation expenses and enable just-in-time delivery.

Availability encompasses a multitude of supply chain factors, including but not limited to the producer's technological prowess and financial stability, the supplier's capacity to ensure high quality through established quality systems and compliance with local or regional regulations, potential economic and political upheavals, and numerous others. Device firms may desire to depend on the infrastructure and knowledge of a global manufacturing partner in order to achieve success in this sector. A sizable manufacturer services and end-to-end design firm specialising in a variety of products generally maintains a robust, global supply chain comprising dependable suppliers, an extensive selection of materials that have undergone stringent testing, and the capacity to proactively oversee and control supplier performance.

Flexible design: The invention of devices is propelled by the design freedoms made possible by materials. In order to enable simpler, more attractive designs, plastics can be consolidated from many components or moulded into complicated shapes on their own. A thermoplastic elastomer that has been overmoulded onto a stiff plastic substrate may offer enhanced traction or a more pleasant feel. Specific composites and polymers can assist architects in the development of devices with thinner walls that are both lightweight and sturdy.

Particularly with the advent of next-generation technologies like conductive polymers, nanopolymers, and shape memory materials, designers may require the assistance of a materials reference database or a knowledgeable partner in order to identify the most suitable candidates from the vast number of plastic chemistries and compounds. Relying on established methods may result in time savings; nevertheless, that might also deprive you of a competitive edge or even an innovative design.

Cost per unit: The expense of materials is a crucial component of every medical gadget. However, in order to determine the full lifespan costs of a plastic or metal, businesses must go beyond its pricing. Costs can be increased or decreased by varying materials' effects on subsequent factors, including but not limited to:

- The quantity needed per device, which is determined by material density, wall thickness, and part consolidation.
- The necessity or non-necessity of additives, as in the case of polymers with moulded-in colour;

Transport expenses associated with shipping weight or distance; and manufacturing simplicity and speed resulting from the material's adaptability for high-volume production or the necessity for secondary procedures. It is critical to evaluate the potential impact of each possible material on the unit cost of a device across its entire lifespan. An initially higher-priced material that exhibits superior performance may ultimately result in cost savings.

Material properties: The process of material selection is predicated on the correspondence between performance capabilities and device requirements. In addition to mechanical attributes such as strength, flexural modulus, and resilience to high temperatures, performance capabilities may also include chemical resistance, including resistance to powerful antibacterial cleansers required to battle hospital-acquired infections. In addition, they incorporate the requirements of typical usage circumstances, such as resistance to impact-induced breaking or cracking. Certain substances possess strength, flexibility, or chemical resistance by nature, whilst others may be strengthened by fillers, glass fibre, or additions.

- Mechanical properties: These are often the primary factors taken into account when selecting a material, since they dictate its appropriateness for the tasks that the component will be required to do. In addition to properties such as yield strength, toughness, and stiffness, environmental conditions of storage and usage (e.g., heat, cold, humidity) and loading circumstances (e.g., compression/tension, continuous/cycled, single or repeated use) assist the engineer in identifying prospective candidates [24].

- Surface properties: Surface qualities, which extend beyond the aesthetics of the medical device and include particular finishing elements influenced by manufacturing limitations (such as scratch resistance or mouldability) or secondary processing, are often established by the design team (e.g., serigraphy, pad printing, and gluing).
- Physical properties: Certain applications rely heavily on broader physical characteristics, such as electrical conductivity, transparency, and density, even if they are not immediately apparent (for example, a build-up of static electricity on a component in the drug path can significantly affect the performance of a dry powder inhaler). Physical qualities may also influence manufacturing alternatives, like ultrasonic or laser welding for assembly or injection moulding.
- Chemical properties: These factors might potentially impact resistance to deterioration caused by lubricants, solvents, moisture, or electromagnetic radiation (e.g., UV light). Additionally, it may give rise to apprehensions regarding vulnerability to sterilisation methods like gamma or ethylene oxide (EtO). Additionally, the chemical composition of a substance dictates whether certain moulding processes, such as co-moulding, may be employed to accomplish distinct characteristics inside a single component or a multi-textured external surface (for "soft-touch" grips), for instance (a clear window within an opaque body or casing, or a compliant sealing element for example).
- Biocompatibility: Critical criteria that must be met by numerous medical devices, particularly drug delivery devices, include the use of materials that are appropriate for both the drug and the user. These materials should be resistant to both the drug and the user's characteristics, and can be applied via skin contact or implantation, whether on a long-term or short-term basis. Intensive testing protocols may be necessary to evaluate extractables and leachables, toxicity, and irritation in the case of unapproved materials, contingent upon the specific use and associated risks. In situations involving medication main packaging or long-term implantation, where permeability to substances such as oxygen or moisture may also be critical features, these concerns may be the primary determinant in the selection of materials and grades.
- Rheological properties: Rheological qualities pertain to the material's potential for undergoing a transformation throughout the manufacturing procedure. In some circumstances, material selection may be required to facilitate injection moulding; for instance, a more fluid material may be preferred.

- Colour: A crucial component of product design that is frequently employed to aid in the development of the visual language required to promote proper operation and to enhance the aesthetic appeal of a gadget. However, the process of selecting colours for medical equipment might pose several challenges. Medical grade is restricted to a specific palette, which restricts the options available for device designs unless financial resources and schedules allow for an extensive approvals procedure. The influence of colour, or the alteration of colour, on moulding tolerances is particularly significant in cases when components have precisely regulated essential dimensions.
- Wear and friction: Particular attention should be given to the material's resistance to wear, lubricity, and friction characteristics if it will come into touch with other surfaces or moving components, such as implants.
- Regulatory compliance: Assure the material complies with standards and criteria required for medical technology manufacture and use, such as those set forth by the FDA or International Organization for Standardization (ISO). This will ensure its safety and efficacy.

Apart from these considerations, specific materials for the above-stated manufacturing techniques (i.e., CNC, injection moulding, and 3D printing) should also be studied. Due to the fact that virtually any material with adequate hardness may be machined, CNC provides an extensive selection of material alternatives. Figure 11.18 shows the decision tree for materials used in CNC based on the material properties.

The primary materials utilised in CNC machining are metals and metal alloys. Low-to-medium batch production as well as the fabrication of unique one-of-a-kind components and prototypes are both possible with metal. The following are the materials which are compatible with CNC machining.

- Aluminium: Aluminium 6061 is the material utilised most frequently in CNC machining. Properties include superior ductility and machinability, with a favourable strength-to-weight ratio. Aluminium alloys have a low density, a favourable strength-to-weight ratio, strong thermal and electrical conductivity, and inherent resistance to corrosion.
- Stainless steel: Its properties include extreme resistance to corrosion and heat, with a high tensile strength. Alloys of stainless steel are rigid, ductile, and resistant to corrosion and wear. It is simple to weld, mill, and polish them.

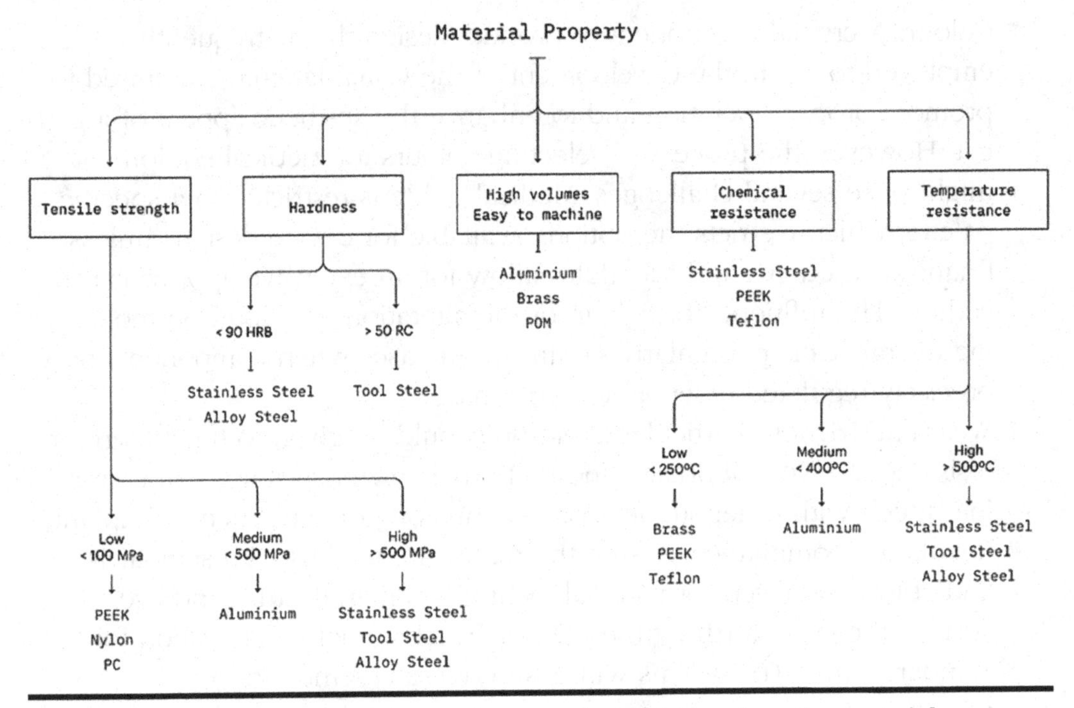

Figure 11.18 Decision tree for materials used in CNC based on the material properties [6].

- Alloy steel: Its properties include extreme strength and toughness; resistant to exhaustion. Alloy steels, apart from carbon, comprise additional alloying elements that contribute to enhanced resistance to fatigue, wear, hardness, and toughness.
- Brass: Some common properties of brass include exceptional electrical conductivity, little friction, and a golden look. Brass is an alloy of metal characterised by exceptional electrical conductivity and machinability. Ideally suited for uses that demand little friction.
- Acrylonitrile butadiene styrene (ABS): Its properties include resistance to impact. ABS is a prevalent thermoplastic renowned for its exceptional mechanical characteristics, including high impact resistance, heat tolerance, and machinability.
- Polyether ether ketone (PEEK): Its common properties include exceptional chemical and heat resistance. PEEK is a thermoplastic for high-performance engineering with superior chemical and mechanical capabilities throughout a broad temperature range.

As far as injection moulding is concerned, all thermoplastics can be injection moulded. Additionally, liquid silicones and some thermosets are compatible with the injection-moulding process. Additionally, their physical qualities may be altered by reinforcing them with fibres, rubber particles, minerals, or flame-retardant chemicals. For instance, the incorporation of fibreglass into the pellets at proportions of 10%, 15%, or 30% can provide components with increased rigidity. The following are the materials which are compatible with injection moulding.

- Polypropylene: The most prevalent plastic for injection moulding. Outstanding chemical resistance. Food-safe grades are offered. Unfit for use in mechanical applications.
- ABS: Low-cost, common thermoplastic with exceptional impact resistance. Susceptible to solvent damage.
- Polyethylene: Weather-resistant and lightweight thermoplastic with excellent impact resistance. Appropriate for outdoor use.
- Polyurethane: An excellent thermoplastic material with exceptional impact resistance, mechanical characteristics, and hardness. Adequate for moulding components with substantial walls.
- Polycarbonate: The material exhibiting the greatest impact resistance. Superior weather resistance, thermal resistance, and hardness. Transparent or tinted formulation.
- Silicone rubber: A thermoset material that is resistant to heat and chemicals and whose shore hardness may be modified. Available in food-safe and medical-grade formulations.

For FDM 3D printing, a wide range of materials is available. Table 11.1 could be considered to select different materials as per the requirement.

Table 11.1 Common Materials Used in FDM 3D Printing

Material Name	*Properties*
Acrylonitrile butadiene styrene (ABS)	Good strength, temperature resistance, warping issues.
Polylactic acid (PLA)	Good visual quality, low impact strength.
Polyethylene terephthalate glycol (PETG)	Could be food safe, good strength.
Thermoplastic polyurethane (TPU)	Flexible, difficult to print.
Polyetherimide (PEI)	Good strength to weight ratio, good fire and chemical resistance, costly.

Table 11.2 Common Materials Used in SLA 3D Printing

Material Name	*Properties*
Standard resin	Smooth surface finish, highly brittle.
High detail resin	Good dimensional accuracy, high cost.
Flexible resin	Rubber like flexibility, lower accuracy.
Dental Resin	Biocompatible, good abrasion resistance, high cost
Tough resin	High toughness and stiffness, lower thermal resistance.

SLA materials are comparatively different from the FDM ones. SLA materials are liquid resins that may be selected according to the intended application of the component, such as a smooth surface finish, resistance to abrasion, or heat resistance. Thermosets, which are SLA materials, are more brittle than FDM-produced materials. Table 11.2 could be considered to select different materials as per the requirement.

References

1. Soori M, Arezoo B, Dastres R. Machine learning and artificial intelligence in CNC machine tools, A review. *Sustain Manuf Serv Econ* 2023;2:100009. https://doi.org/10.1016/J.SMSE.2023.100009.
2. Medical Device Manufacturing n.d. https://www.meddeviceonline.com/resource/medical-device-manufacturing (accessed December 28, 2023).
3. Wang Z, Rahman M. High-speed machining. *Compr Mater Process* 2014;11:221–53. https://doi.org/10.1016/B978-0-08-096532-1.01113-4.
4. Medical device manufacturing - everything you need to know n.d. https://www.jamesspring.com/news/what-is-medical-device-manufacturing/ (accessed December 28, 2023).
5. What is CNC machining? | Hubs n.d. https://www.hubs.com/knowledge-base/cnc-machining-manufacturing-technology-explained/ (accessed December 28, 2023).
6. CNC machining: The manufacturing & design guide | Hubs n.d. https://www.hubs.com/guides/cnc-machining/#basics (accessed December 28, 2023).
7. How to prepare a technical drawing for CNC machining | Hubs n.d. https://www.hubs.com/knowledge-base/how-prepare-technical-drawing-cnc-machining/ (accessed December 28, 2023).
8. What is CNC milling? | Hubs n.d. https://www.hubs.com/knowledge-base/what-is-cnc-milling/ (accessed December 28, 2023).
9. Injection molding: The manufacturing & design guide n.d. https://www.hubs.com/guides/injection-molding/#the-basics (accessed December 28, 2023).

10. What is FDM (fused deposition modeling) 3D printing? | Hubs n.d. https://www.hubs.com/knowledge-base/what-is-fdm-3d-printing/ (accessed December 28, 2023).
11. Bhuvanesh Kumar M, Sathiya P. Methods and materials for additive manufacturing: A critical review on advancements and challenges. *Thin-Walled Struct* 2021;159:107228. https://doi.org/10.1016/J.TWS.2020.107228.
12. Mobarak MH, Islam MA, Hossain N, Al Mahmud MZ, Rayhan MT, Nishi NJ, et al. Recent advances of additive manufacturing in implant fabrication – A review. *Appl Surf Sci Adv* 2023;18:100462. https://doi.org/10.1016/J.APSADV.2023.100462.
13. Boretti A, Castelletto S. A perspective on 3D printing of silicon carbide. *J Eur Ceram Soc* 2024;44:1351–60. https://doi.org/10.1016/J.JEURCERAMSOC.2023.10.041.
14. Chatterjee S, Gupta S, Chanda A. Barefoot slip risk assessment of Indian manufactured ceramic flooring tiles. *Mater Today Proc* 2022;62:3699–706. https://doi.org/10.1016/J.MATPR.2022.04.428.
15. Gupta S, Chanda A. Biomechanical modeling of footwear-fluid-floor interaction during slips. *J Biomech* 2023;156:111690. https://doi.org/10.1016/J.JBIOMECH.2023.111690.
16. Gupta S, Malviya A, Chatterjee S, Chanda A. Development of a portable device for surface traction characterization at the shoe-floor interface. *Surfaces* 2022;5:504–20. https://doi.org/10.3390/SURFACES5040036.
17. Gupta S, Jayaraman R, Sidhu SS, Malviya A, Chatterjee S, Chhikara K, et al. Diabot: development of a diabetic foot pressure tracking device 2023;6:32–47. https://doi.org/10.3390/J6010003.
18. Kholgh Eshkalak S, Rezvani Ghomi E, Dai Y, Choudhury D, Ramakrishna S. The role of three-dimensional printing in healthcare and medicine. *Mater Des* 2020;194:108940. https://doi.org/10.1016/J.MATDES.2020.108940.
19. Aquino Monteiro S, Scheid C, Deon M, Merib J. Fundamentals, recent applications, and perspectives of 3D printing in sample preparation approaches. *Microchem J* 2023;195:109385. https://doi.org/10.1016/J.MICROC.2023.109385.
20. Bao W. Resolving warping in 3D printing of thermoplastic parts via heterostructure brim. *Mater Lett* 2023;351:135032. https://doi.org/10.1016/J.MATLET.2023.135032.
21. Quan H, Zhang T, Xu H, Luo S, Nie J, Zhu X. Photo-curing 3D printing technique and its challenges. *Bioact Mater* 2020;5:110–5. https://doi.org/10.1016/J.BIOACTMAT.2019.12.003.
22. What is SLA 3D printing? | Hubs n.d. https://www.hubs.com/knowledge-base/what-is-sla-3d-printing/ (accessed December 28, 2023).
23. 5 Innovative use cases for 3d printing in medicine | Formlabs n.d. https://formlabs.com/asia/blog/3d-printing-in-medicine-healthcare/ (accessed December 28, 2023).
24. Liu K, Takasu K, Jiang J, Zu K, Gao W. Mechanical properties of 3D printed concrete components: A review. *Dev Built Environ* 2023;16:100292. https://doi.org/10.1016/J.DIBE.2023.100292.

Chapter 12

Prototype Functionalisation Using Electronics and Instrumentation

12.1 Introduction

The utilisation of electronic equipment is vital for medical treatments [1]. The progress made in technology has resulted in a heightened need for sophisticated medical equipment. Medical electronics are necessary for the execution of procedures such as operations, therapies, and the diagnosis of health issues [2]. Medicine is currently experiencing rapid growth in a society where individuals prioritise superior health services. Medical electronics are crucial in facilitating the efficient functioning of the medical sector. Additionally, they are the apparatus that renders surgeons and physicians impotent. Medical electronics are specifically engineered to identify and resolve health-related issues. In addition, artificial gadgets can be classified as medical electronics. Medical electronics find application in a variety of imaging technologies and medical treatments, including magnetic resonance imaging (MRI).

Electronics for medical purposes are fundamental to health systems. They are gadgets particularly developed to fulfil medicinal objectives. These electronic technologies assist health practitioners to detect medical conditions and treat patients. Also, medical electronics contribute in increasing the quality of life. They assist to prevent, diagnose, and treat health disorders. According to the US Food and Drug Administration (FDA), medical

DOI: 10.4324/9781003475309-16

electronics are examined based on the danger to patients. High risk items need greater clinical proof than lower risk medicinal products. Also, the FDA only authorises items that have satisfied its standards to be accessible on the US market.

In vitro diagnostic gadgets are also classified as medical electronic devices [3]. Additionally, test samples from the body, such as blood, tissue, and biological fluids, are carried out with the aid of this equipment. Defibrillators, hearing aids, and diagnostic equipment for tracking, logging, and measuring bodily processes including heartbeat and brain waves are a few examples of medical electronic devices. For medical electronics to work, printed circuit boards with excellent performance and dependability are needed. As medical electronics are primarily used to save lives, it is also essential to incorporate high-quality boards into their manufacturing process. As a result, the functioning and quality of the circuit boards employed for these devices are highly important [4]. Medical electronic gadgets come in several varieties. These gadgets aid in the treatment of internal medical disorders.

Monitoring devices: Medical monitoring equipment is frequently utilised in a variety of diagnostic and therapeutic processes. Additionally, they support the thorough and effective execution of these operations. Heart rate and blood pressure monitors, electromyography (EMG) activity systems, flight rate and dispensing systems, X-ray computed tomography, and body temperature monitors are a few examples of medical monitoring devices.

Medical diagnostic devices: Medical image acquisition, detection, and display are common uses for these medical electronics. They also offer important details about the human body. Ultrasound equipment, computed tomography (CT) scanners, and MRI are a few types of medical diagnostic gadgets. A CT scanner makes use of X-rays, specialised lenses, and computer algorithms to produce images of a patient's interior organs. Sound waves produced by ultrasound equipment reverberate off human tissue. Additionally, this device displays and converts the picture of a structure into an image.

Implantable medical devices: Implantable devices aid in the replacement of the body's damaged organs. Furthermore, they are inserted straight into the patient's body. They take the place of any damaged organ. The most popular implantable devices include artificial kidneys, cochlear implants, and heart pacemakers. These devices' essential component is a medical printed circuit board (PCB).

Based mostly on dangers, medical electronics are divided into three types. Instead of using paper-based systems, the majority of top medical device manufacturers now use computerised quality management systems. It's also critical to comprehend the three risk-based medical electronics classes. The following are the different classes of medical devices.

Class I: Medical devices classified as class 1 are not intended to support or preserve life. The FDA states that there may not be an excessive risk of injury from these devices [5]. Class I devices make up more than 47% of all medical devices on the market. Additionally, there aren't many regulations or limitations applicable to class I equipment. Nonetheless, there are some general rules that these devices must follow, such as those regarding branding, adulteration, and device registration.

Class II: Compared to class I, this category of medical devices carries a higher risk. The FDA claims that generic controls for class II devices are insufficient to provide a reasonable assurance of the device's efficacy and safety. Additionally, class II devices have different controls depending on the device. Common controls include, however, post-market surveillance, particular labelling requirements, and device performance.

Class III: Medical equipment classified as class III are those that maintain or support life. Furthermore, they are typically implanted and may come with an excessive risk of disease or harm. Class III devices include heart pacemakers and break implants. Class III medical electronic devices make up about 10% of the total. These gadgets are therefore heavily regulated. Class III devices are typically used for state-of-the-art medical applications. Class III devices are also the most critical and have high risk.

Manufacturing medical electronics demands attention to detail, accuracy, precision, and quality [6]. Medical device producers also need to consider accuracy and quality when creating these gadgets. These tools were created specifically to save lives. As a result, the manufacture of medical devices must be taken into account. The following are some factors that affect the medical electronics manufacturing.

Safety: This is a crucial matter. It is the most important issue that manufacturers of medical devices need to take into account. The usage of medical gadgets must be safe. They ought to be secure for the environment and for people as well. Additionally, these gadgets need to be designed by manufacturers to avoid any kind of shock. Medical implanted devices, for example, require very high precision.

Compliance: The manufacturing of medical devices needs to adhere to various regulations. Medical devices must also adhere to tight guidelines.

Medical gadgets are subject to particular criteria. These specifications assess the accuracy and calibre of medical equipment.

Precision: When it comes to the fabrication of medical electronics, precision is crucial. It is imperative for medical device makers to ascertain the precision of their products. Furthermore, since the primary purpose of medical equipment is to save lives, precision is extremely important. These gadgets should also be resistant to hazardous situations.

Lifespan: The longevity of medical equipment must be taken into account during production. Medical PCBs are one of the elements that affect how long medical gadgets last. For medicinal electronic equipment to be durable and of high quality, its lifespan is essential.

Marginal cases: The marginal cases of a medical device's application environment must be taken into account by medical device manufacturers. An entrepreneur ought to be aware of the ideal temperature and other environmental conditions for the gadget to function.

Medical electronics manufacturing can be a challenging procedure. The medical electronics sector is a multifaceted industry. These technologies are also constantly in great demand. As a result, this provides hope for the future. Manufacturers of medical devices do, however, encounter certain difficulties in the market. For manufacturers to succeed and enhance public health, these obstacles must be addressed. The following are the major challenges faced by the electronic manufacturing industries for new product or medical device development.

High healthcare costs: The medical industry faces a challenge from rising healthcare expenses. Many people worldwide lack access to affordable, high-quality healthcare. Manufacturers of medical devices are under pressure to lower the price of their goods. Because of this, producers are choosing to produce disposable goods rather than durable ones.

Counterfeit: The market will see a rise in fake goods as the need for medical electronic gadgets grows. Manufacturers of legitimate devices may see a shift in revenue due to fraudulent medical products. Additionally, people who use them may suffer injury from them. Integrating a system to verify the legitimacy of their items is the most effective technique to get rid of counterfeits.

Expensive and slow research: Research and development play a crucial role in the production of medical devices. On the other hand, doing the required clinical trials is costly. Additionally, this process could be extremely slow, which would prolong the time to market. Long-term profitability may be hampered by this. Using robotic process automation and cloud technology is one way to overcome this obstacle.

Changing supply chain: During the COVID-19 pandemic, there were increased difficulties for the medical device manufacturing sector. Various nations implemented different lockdown measures during the outbreak. Supply chains were disrupted as a result. Additionally, this resulted in further delivery delays and rising material costs. This can be resolved by integrating Internet of Things (IoT) technologies into the medical supply chain.

Connecting and communicating with patients is made easier for healthcare providers by electronic gadgets. As medical professionals search for methods to enhance patients' health, medical electronics have grown to be an essential component of the healthcare sector. In the medical industry, electronics are used in a variety of ways. Some of these include:

Brain wave machine: A brain wave machine is an essential piece of medical equipment for recording the electrical activity of the scalp. It functions by releasing endogenous neurons in the brain. Additionally, this device processes the information obtained from the electrodes that are applied to the scalp. The brain wave machine shows its findings on a screen. Additionally, this device aids in the treatment of mental illnesses like insomnia, brain death, and mental instability. It is very helpful in emergency rooms.

Defibrillator: A defibrillator is an essential tool in an emergency. It is mostly applied in cases of emergency, like heart attacks. Moreover, this medical gadget has an impact on cardiac arrhythmia, ventricular fibrillation, and pulseless ventricular tachycardia. The heart receives an electric shock from the defibrillator to function. The heart's muscles then depolarise as a result, reproducing the electrical pulse's regular conduction of heat.

MRI: The interior components of the human body are examined by the MRI machine. Radiology makes extensive use of this apparatus. Moreover, an intense magnetic field is used by MRI machines to make images of the body. MRI machines are diagnostic tools that can be used to diagnose and treat a variety of medical conditions. Moreover, radiation is not integrated into MRI. It is therefore far better than a CT scanner.

Blood gas analyser: The best use of electronics in the medical field is the blood gas analyser. It determines the blood pressure of chemicals such as carbon monoxide and oxygen. Medical professionals can use result analysis to identify any blood disease. This machine uses a chemical device strip to gather blood samples. Particle selective electrodes are a feature of the chemical device strip.

Electronic cardiac monitor: The pressure waveforms of the heart system are displayed with the aid of this medical electronic equipment. To acquire an electrocardiogram (ECG) of the cardiac system, medical professionals place certain electrodes on the patient's body. This monitor additionally looks for any abnormal heart activity. It is frequently utilised in medical care, particularly in surgery.

Chronic health issues such as diabetes and high blood pressure can be remotely monitored thanks to medical electronic gadgets. These gadgets can also send data to caregivers from a patient's home. Additionally, medical electronics facilitate patient connections and communication for healthcare professionals. Here's just one example of how medical electronics are transforming people's lives and the wider world. Medical devices also track, identify, and treat human health issues. As a result, these gadgets are typically seen as essential components of health systems. By safely preventing, diagnosing, and treating illnesses and disorders, they enhance the quality of health.

There are several ways that medical gadgets lower healthcare expenses. One way that healthcare costs have been managed is through the incorporation of electronic health records. When medical professionals have access to fast and precise sickness diagnosis, medical technologies can also lower healthcare expenditures. The time spent on illness diagnosis and treatment can be reduced with the use of medical technology. These gadgets typically deliver results fast. Additionally, medical technologies allow healthcare professionals to concentrate more on critical tasks [7]. Moreover, hospital workflow is streamlined by these technologies. Medical equipment both improve results and facilitate prompt intervention. Medical electronics both save lives and enhance health. These gadgets support sustainable healthcare practices as well. Both individuals and the whole healthcare systems can benefit from the work of the medical business. Making wise clinical judgements is aided by accurate diagnostic information for medical professionals.

12.2 Microcontroller Boards and Common Healthcare Sensors for Prototyping

12.2.1 Microcontroller Boards

Any healthcare or everyday device which requires automated working involves a part which controls all the movements or algorithms, known as a microcontroller board. To understand it better, just like a brain is the primary controlling unit of a human being or an animal, a microcontroller unit acts as the brain of a device or a machine. A microcontroller is a standalone electronic device that consists of an integrated circuit (IC) package housing the CPU, ROM, RAM, clock circuitry, and I/O circuitry. Unlike conventional microprocessors, which require auxiliary chips to function, microcontrollers operate without them.

The signal processing tasks of a comparator, limiter, rectifier, integrator, differentiator, logarithmic amplifier, active filter, and phase-sensitive demodulator are frequently carried out by a microcontroller in software, replacing analogue circuitry. Another name for a microcontroller is a digital microcomputer. When a microcontroller is used, there are typically fewer integrated circuit packages, which reduces complexity. This has the advantage of cost and space. Furthermore, the self-calibration and error detection capabilities of microcontrollers improve the biomedical instrument's reliability.

Microcontrollers can be used to automatically sequence events, self-calibrate biomedical measurement equipment, and give a simple means of entering patient data, such as height and weight, to calculate expected or normal performance. The fundamental design of the microcontroller system enables all of these features. Due to the on-chip analogue-to-digital converters, microcontrollers—standalone devices for use in data collecting and control—can be directly utilised in biomedical instrumentation. Additionally, we offer microcontrollers with chip communication controllers, which are intended for use in applications where communication between dispersed nodes and local intelligence at remote nodes are necessary. Microcontrollers are widely used in portable medical equipment due to their low maintenance costs, reduced hardware design and board density, and ease of use. Figure 12.1 shows a typical microcontroller board.

An Integrated Development Environment (IDE) and a development board are two essential tools for beginning embedded development. A PCB comprising hardware and circuitry intended to make experimenting with a particular microcontroller board feature easier is called a microcontroller development board. A CPU, RAM, chipset, and on-board peripherals with debugging features, such as an LCD, keypad, USB, serial port, ADC, RTC, motor driver ICs, SD card slot, Ethernet, etc., are all coupled with the development boards.

The bus type, processor type, memory, number of ports, port type, and operating system are among the specifications of microcontroller boards. Programmes for embedded devices, including various controllers, home appliances, robots, kiosks, point-of-sale (PoS) terminals, and information appliances, are evaluated using these. We will talk about the differences between several development boards worldwide in this section. Every one of these has advantages and disadvantages of their own, and certain development platforms are more popular for specific projects than others.

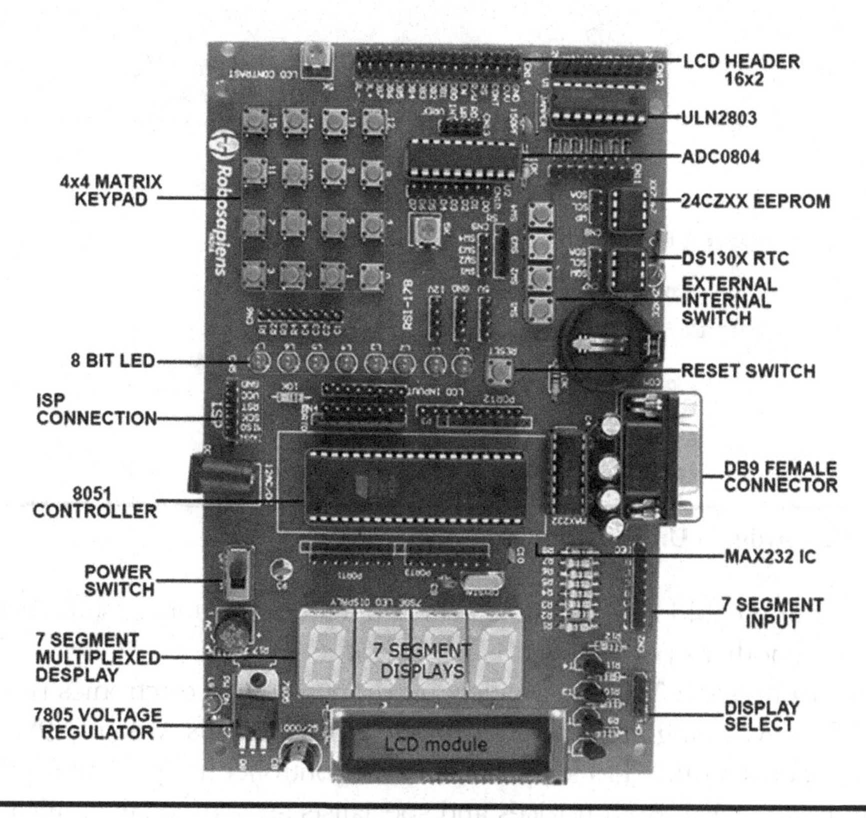

Figure 12.1 A typical microcontroller board [8].

The single board microcontroller, or microcontroller development board, is another name for it. These days, creating a single board microcontroller development kit is incredibly easy and affordable. There are a tonne of open-source development environments (DEs) available for creating microcontroller boards and real-time applications. The following boards are common and could be used for basic electronics proofing or prototyping for healthcare devices.

Do-it-yourself (DIY) based microcontrollers: DIY microcontroller boards could be created at home by anyone. To do this, one will need all the individual electrical and electronic parts, such as component bases, serial ports, LCD modules, keyboards, touchpads, and microcontrollers (Atmel, ARM, and MSP, among others). It will be necessary to properly solder each of these parts to the PCB. After completing the hardware setup, an appropriate IDE could be selected to programme the microcontroller and create the necessary application. Here, a generic and most common microcontroller (i.e., 8051) could be used. The general-purpose 8051 microcontroller is used to create entry-level applications. Examples include industrial and healthcare temperature control systems, automated light intensity control systems, and data acquisition systems. However, it may require good expertise in

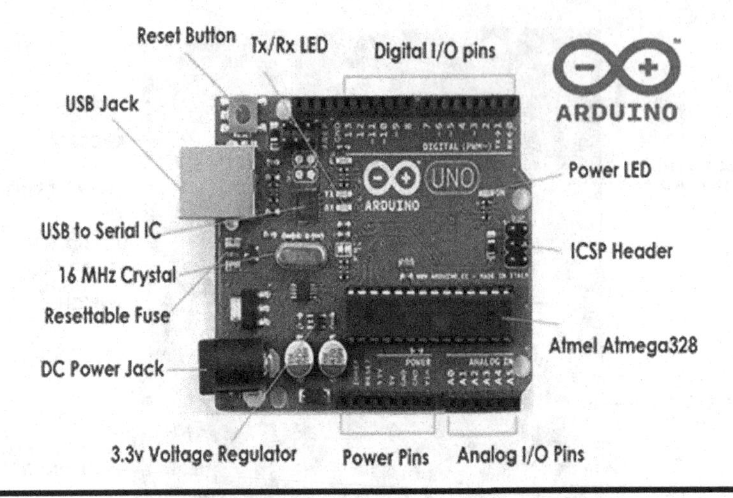

Figure 12.2 Arduino Uno board [8].

electronics and could lead to high costs and lead time. Hence, standardised boards are another option which is explained as follows.

Arduino boards: The most widely used open-source electronics prototyping tool for developing interactive electronic applications is Arduino [9]. All the components required to support the microcontroller are present on the Arduino UNO board. Both novices and specialists are extremely familiar with the Arduino UNO microcontroller board. It is arguably among the earliest development boards based on microcontrollers. The most straightforward and robust prototyping environment built around the ATmega328P microprocessor is the Arduino UNO R3 (Figure 12.2). Its basic features include a 32 kB flash memory, it is operated at 5 volts and the input voltage can range from 7 to 12 volts. It has 14 input and output pins which could be used to connected different sensors and other attachments. Its open-source IDE for creating sketches and its easily comprehensible, "C" language syntax are the main factors contributing to its widespread use. There are other Arduino boards too. If a device requires additional sensor supports and extra pins, other variants such as Arduino Mega or Duo could be used as the basic working is similar to Uno.

Raspberry Pi development board: Like a credit card computer, the Raspberry Pi development board is compact (Figure 12.3). It's simple to connect the Raspberry Pi to a TV, computer, or monitor. It also makes use of a normal mouse and keyboard. It is essential for configuring surveillance cameras and digital media systems, even for non-technical individuals. Without a doubt, the Raspberry Pi 3 is the most capable and reasonably priced computer platform. The newly released Raspberry Pi 3 included a 1.2 GHz, 64-bit quad-core CPU, 802.11n Wireless LAN, Bluetooth 4.1, Bluetooth

Figure 12.3 Raspberry Pi board [8].

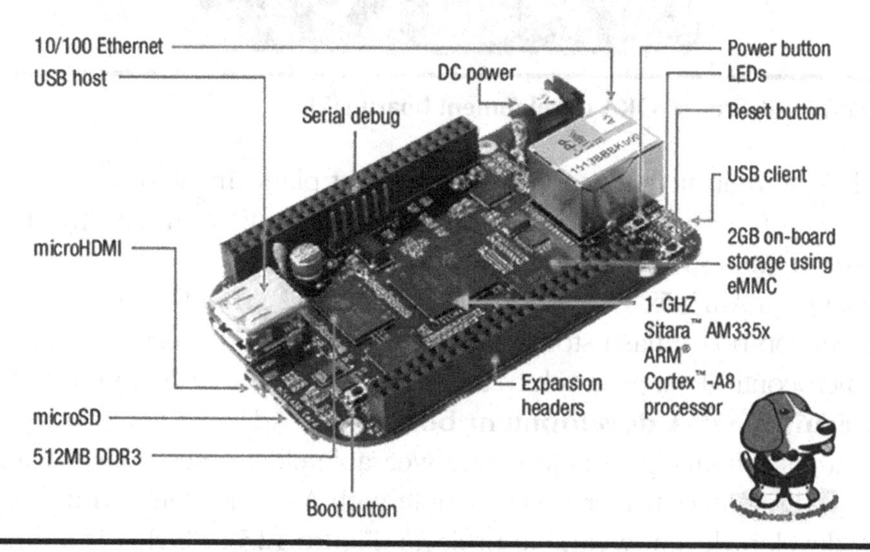

Figure 12.4 BeagleBoard development board [8].

Low Energy (BLE), 1GB RAM, 4 USB ports, 40 general purpose input-output (GPIO) pins, full HDMI port, combined 3.5 mm audio jack and composite video, etc. [10]. Running on a customised version of Debian Linux known as Raspbian, Raspberry Pi can install a variety of software, such as Python, Java, the LAMP stack, Node.js, and many more.

BeagleBone Black development board: BeagleBone Black is a well-known open-source computing platform (Figure 12.4). It now has integrated wireless networking capabilities [11]. BeagleBone Black Wireless, created in CadSoft Eagle and made possible by a collaboration with Octavo Systems, is the most user-friendly credit card-sized IoT Linux computer on the market. For those creating embedded applications, BeagleBone Black is an

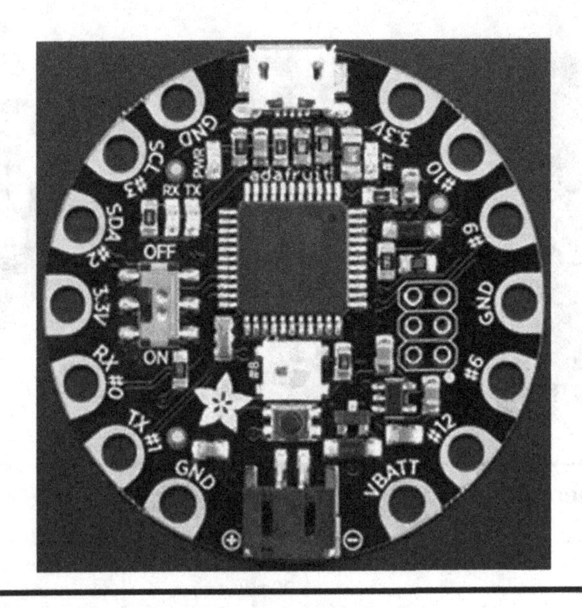

Figure 12.5 AdaFruit FLORA development board [12].

affordable, community-supported development platform. With just a single USB cable, you may begin developing in less than five minutes after installing Linux in ten seconds during the booting process.

It includes a AM335x 1 GHz ARM Cortex-A8, 512 MB DDR3 RAM, 2 GB 8-bit eMMC on-board flash storage, NEON floating-point accelerator, 2× PRU 32-bit microcontrollers, 3Dgraphics accelerator, HDMI and 2× 46 pin headers.

AdaFruit FLORA development board: The AdaFruit FLORA development board's primary goal is to create wearable electronics. This sewable, disk-shaped microcontroller is compatible with Arduino IDE and usually used to develop decent wearable projects (Figure 12.5). FLORA is a little (1.75″ diameter, weighing 4.4 grams) board and has the finest sensors, GPS modules, chainable LED NeoPixels, and stainless-steel threads are all part of the FLORA family and make ideal add-ons for the FLORA core board [12]. Due to its polarised connector and protective diodes, FLORA is incredibly beginner-friendly and is difficult to ruin by connecting a battery backwards. Because of the on-board regulator, damage or tearing won't occur even when a 9 V battery is connected.

ESP32 development board: A low-cost System on Chip (SoC) microcontroller from Espressif Systems, the company behind the well-known ESP8266 SoC, is called the ESP32. It is a single-core and dual-core 32-bit Tensilica Xtensa LX6 Microprocessor with integrated Wi-Fi and Bluetooth that replaces the ESP8266 SoC [13]. Like the ESP8266, the ESP32 has the advantage of having integrated RF components such as an RF balun, filters,

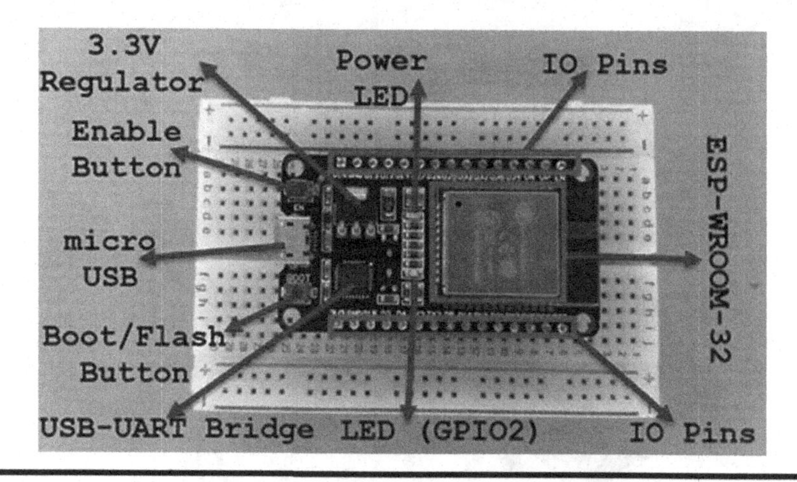

Figure 12.6 ESP32 development board [13].

an antenna switch, a low noise receive amplifier, and a power amplifier. As a result, building hardware using the ESP32 is rather simple because very few external components are needed. In the past, ESP32 was used by Gupta et al. [14] to measure the plantar pressures for effective mapping of diabetic foot ulcers. Figure 12.6 shows a ESP32 development board.

12.2.2 Common Healthcare Sensors

With the opportunities from developing technology, the subject of patient health monitoring is significant and expanding quickly these days. Several open source and readily available sensors are available in the market which are easily compatible with each of the above explained boards. The following will explain some of the commonly used biometric sensors compatible with common boards.

Optical fingerprint reader: This optical fingerprint sensor that is compatible with Arduino makes it easy to incorporate fingerprint identification into the projects. There are other techniques too, for the evaluation of fingerprints [15]. The gadget just captures optical pictures of the fingerprint, as the name suggests. After that, it analyses the image's bright and dark areas to identify unique patterns. Simply connect the sensor to the Arduino via TTL Serial, enrol in any online libraries that are available, and use the sensor to read the fingerprint [16]. Now that the module is configured and programmed, it can be used to trigger additional devices, such device locks etc. Figure 12.7 shoes a typical optical fingerprint sensor.

Pulse sensor: The pulse sensor (Figure 12.8) is a plug-and-play module that makes it simple to include real-time heart rate data into any healthcare

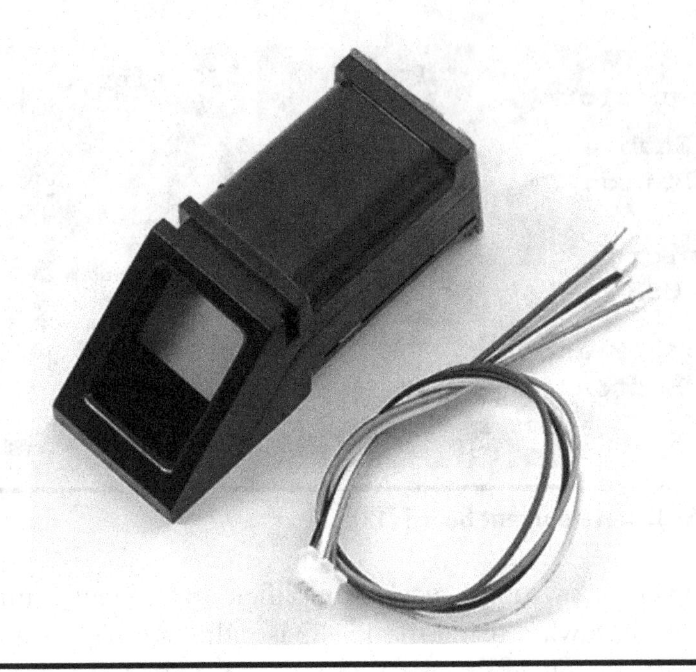

Figure 12.7 Optical fingerprint reader [17].

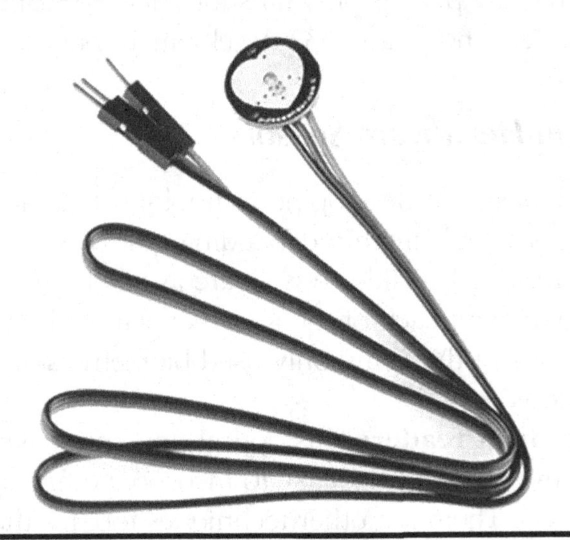

Figure 12.8 Pulse sensor [17].

device [18]. To monitor the change in light generated by the capillary blood vessel expansion, it includes an optical sensor, an amplifier, and a noise-suppression circuit. An added advantage is that only three pins (i.e., an analogue data pin, VCC, and GND) are required for setup, making the sensor exceedingly simple to use. The setup can read the heart rate by just placing the finger on the sensor's front-facing heart logo, plugging it into the microcontroller board using jumper cables.

Figure 12.9 MAX30102 sensor module.

MAX30102 pulse oximeter and heart rate sensor module: One of the best methods to incorporate health monitoring into the healthcare device is with the MAX30102 sensor (Figure 12.9), which combines a heart rate monitor and pulse oximeter into a single device [19]. This small sensor gauges the blood's oxygen saturation by measuring the amount of red and infrared light reflected by the blood. The heart rate can also be determined by examining the time series response of these reflected infrared and red photons. The MAX30102 Pulse and Heart Rate Sensor has a photodetector that can detect both visible and infrared light in addition to two LEDs (red and infrared). It is very low power and connects to the microcontroller via an I2C digital interface, which makes it perfect for wearables with projects and other mobile applications.

Heart monitoring sensor: The affordable AD8232 Heart Rate Monitor (Figure 12.10) captures ECGs while on the go. It allows an eye to be kept on the heartbeat by measuring the electrical changes caused by cardiac motions. Three electrodes that are positioned on the left, right, and right leg can be used to record these electrical signals [20]. The signals are then extracted, amplified, and filtered by the AD8232 sensor to create a voltage that Arduino and other microcontrollers can read. The AD8232 Heart Rate Monitor can offer some useful information for basic heart monitoring even though it isn't as powerful as the ECGs seen in hospitals. Convert it into a portable ECG to monitor anyone's heart rate quickly and improve bike rides or other physical activities.

Galvanic skin response sensor module: To gauge someone's emotional spikes, a device called a galvanic skin response (GSR) sensor could be used [21]. The perspiration glands usually become overactive when anyone is feeling intense emotions. This causes the body to perspire more, which alters the electrical conductivity of the skin. The GSR sensor uses two electrodes that

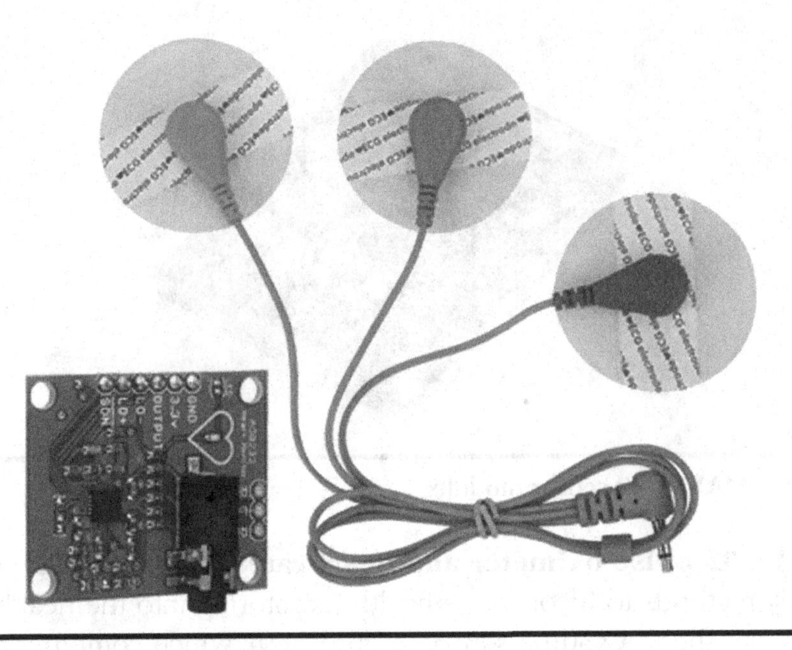

Figure 12.10 AD8232 heart monitor module.

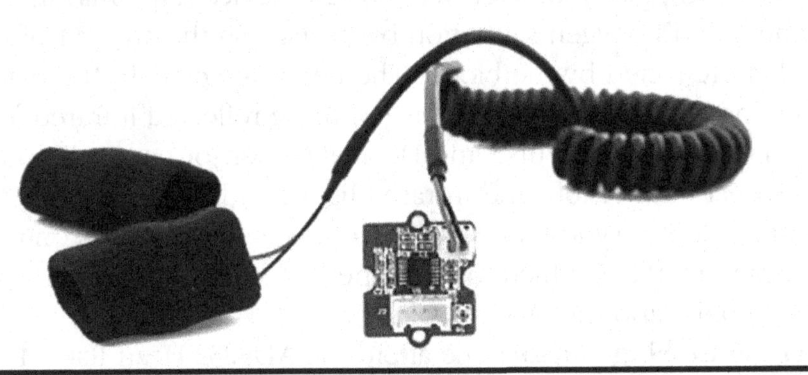

Figure 12.11 Galvanic skin response sensor module [17].

are affixed to the fingertips to identify this sudden shift in electrical conductivity. After interpreting this shift, it finally produces an analogue signal that the Arduino can read. Essentially, the GSR sensor can identify changes in an individual's physiological condition, which makes lie detectors one of the most popular (and entertaining) uses for the sensor. It's also a fascinating supplement to other initiatives that deal with emotions, like sleep quality monitoring. Figure 12.11 shows a typical setup for galvanic skin sensor.

MQ-3 Alcohol sensor: The MQ3 alcohol sensor (Figure 12.12) can be used to include an alcohol sniffer. This sensor detects the amount of alcohol in the air and determines its concentration. Therefore, utilising the MQ3 alcohol sensor could be of great choice in developing a device to detect alcohol or breath analyser. To get the MQ3 sensor to interface with the Arduino, it is

Figure 12.12 MQ-3 Alcohol sensor module [17].

Figure 12.13 Myoware muscle sensor [17].

needed to create a simple drive circuit. Connect the sensor to the Arduino, add a load resistor, and supply 5 V to power it. Next, depending on the amount of alcohol in the air, the sensor outputs a voltage that varies.

Myoware muscle sensor: With this clever sensor, anyone can simply flex their arm, bend their knee, or even clench their fist to begin working on the device. This is achieved by EMG, in which the electrical activity of muscles is measured and amplified by the muscle sensor [22]. The electrical signal from the muscle is then converted into a variable voltage, which you can read using the analogue input pin on the Arduino. Additionally, using the muscle sensor is really easy. All it takes to begin triggering almost anything is a few biomedical sensor pads affixed to the preferred muscle. Figure 12.13 shows the Myoware muscle activity detecting sensor.

References

1. Goncu-Berk G, Topcuoglu N. A healthcare wearable for chronic pain management. *Design of a Smart Glove for Rheumatoid Arthritis. Des J* 2017;20 Supl:S1978–88. https://doi.org/10.1080/14606925.2017.1352717.
2. How medical electronics is contributing to development in the medical industry - RAYPCB n.d. https://www.raypcb.com/medical-electronics/ (accessed January 3, 2024).
3. Weightman APH, Preston N, Holt R, Allsop M, Levesley M, Bhakta B. Engaging children in healthcare technology design: developing rehabilitation technology for children with cerebral palsy. *J Eng Des* 2010;21:579–600. https://doi.org/10.1080/09544820802441092.
4. The impact of electronics for medicine in the healthcare sector | MADES n.d. https://mades.es/blog/impact-of-electronics-for-medicine/ (accessed January 3, 2024).
5. Food and Drug Administration. The 510(k) program: evaluating substantial equivalence in premarket notifications [510(k)] guidance for industry and food and drug administration staff 2014 https://www.fda.gov/regulatory-information/search-fda-guidance-documents/510k-program-evaluating-substantial-equivalence-premarket-notifications-510k
6. The importance of electronics in modern medicine | Sellectronics n.d. https://www.sellectronics.co.uk/blog/the-importance-of-electronics-in-modern-medicine/ (accessed January 3, 2024).
7. Five ways electronics are supporting new innovations in the medical sector n.d. https://www.nsmedicaldevices.com/analysis/electronics-in-medical-innovations/ (accessed January 3, 2024).
8. Various kinds of microcontroller boards with applications n.d. https://www.elprocus.com/different-types-of-microcontroller-boards/ (accessed January 3, 2024).
9. Arduino Uno Rev3 — Arduino online shop n.d. https://store-usa.arduino.cc/products/arduino-uno-rev3?selectedStore=us (accessed January 3, 2024).
10. Buy a Raspberry Pi 5 – Raspberry Pi n.d. https://www.raspberrypi.com/products/raspberry-pi-5/ (accessed January 3, 2024).
11. BeagleBone® Black - BeagleBoard n.d. https://www.beagleboard.org/boards/beaglebone-black (accessed January 3, 2024).
12. FLORA - Wearable electronic platform: Arduino-compatible [v3] : ID 659 : $14.95 : Adafruit industries, unique & fun DIY electronics and kits n.d. https://www.adafruit.com/product/659 (accessed January 3, 2024).
13. Introduction to ESP32 | Specifications, ESP32 DevKit Board, Layout n.d. https://www.electronicshub.org/getting-started-with-esp32/ (accessed January 3, 2024).
14. Gupta S, Jayaraman R, Sidhu SS, Malviya A, Chatterjee S, Chhikara K, et al. Diabot: development of a diabetic foot pressure tracking device 2023;6:32–47. https://doi.org/10.3390/J6010003.

15. Chong AMS, Yeo BC, Lim WS. Integration of UWB RSS to Wi-Fi RSS fingerprinting-based indoor positioning system. *Cogent Eng* 2022;9. https://doi.org/10.1080/23311916.2022.2087364.
16. 8 Arduino-compatible biometric sensors for Hobbyists n.d. https://www.makeuseof.com/arduino-compatible-biometric-sensors-for-hobbyists/ (accessed January 3, 2024).
17. 9 Biometric sensors Arduino compatible | Random Nerd Tutorials n.d. https://randomnerdtutorials.com/9-biometric-sensors-arduino-compatible/ (accessed January 3, 2024).
18. Das S, Pal S, Mitra M. Arduino-based noise robust online heart-rate detection. *J Med Eng Technol* 2017;41:170–8. https://doi.org/10.1080/03091902.2016.1271044.
19. Nemomssa HD, Raj H. Development of low-cost and portable pulse Oximeter device with improved accuracy and accessibility. *Med Devices Evid Res* 2022;15:121–9. https://doi.org/10.2147/MDER.S366053.
20. Sampath A, Sumithira TR. Sparse based recurrent neural network long short term memory (rnn-lstm) model for the classification of ecg signals. *Appl Artif Intell* 2022;36. https://doi.org/10.1080/08839514.2021.2018183.
21. Barber TX, Coules J. Electrical skin conductance and galvanic skin response during "hypnosis." *Int J Clin Exp Hypn* 1959;7:79–92. https://doi.org/10.1080/00207145908415811/ASSET//CMS/ASSET/2521D1A2-B50E-41C3-BD46-EED0CA549E50/00207145908415811.FP.PNG.
22. 2023 RESNA conference: move to the BeAT of innovation. *Assist Technol* 2024;36:82–98. https://doi.org/10.1080/10400435.2024.2273169.

TESTING

Chapter 13

Test Marketing

13.1 Test Marketing

After completing the prototyping or overall design of the proposed health-care product or service, the product or service may be tested in accordance with the scenarios present in the real market. For this, test marketing is performed. The technique of testing to find out how customers react to a product or marketing campaign is known as test marketing. Test marketing is only done on a small scale and with a select group of individuals [1]. Furthermore, it's possible that none of the process participants are aware of the test or experiment. Test marketing is the most effective technique for refining a product and goes beyond just raising consumer knowledge of it. The market test is also used to develop a brand's reputation and clientele [2].

The kind of manufacturing process at play determines how relevant test marketing is to a certain business. For example, the confectionery industry can often produce enough of the new product for test market purposes with minor changes to existing plants and layouts; however, if the product is to be launched nationally, it might be necessary to build a completely new plant, which would require a much larger investment. It goes without saying that test markets are inappropriate in businesses like automobiles and aircraft where the technology demands the same level of expenditure to produce one unit as it does a thousand.

Therefore, the amount of risk and expenditure that the marketing choice implies must be considered when determining if test marketing is a pertinent component of marketing strategy [3]. Usually, the cost of the equipment and machinery required to make the new product is the most obvious

DOI: 10.4324/9781003475309-18

aspect of the risk. However, there are additional significant costs associated with new products as well. These include advertising costs, the time sales representatives must spend pitching the new product, the company's reputation and goodwill at wholesale and retail levels, the shelf space the new product requires in stores, and the production and rework issues that come with starting a new line. Test marketing is ideal if it allows for a significant portion of these crucial expenses to be postponed until a trustworthy estimate of national sales based on the product's real experience in the marketplace can be generated.

The marketing firm can profit from two key advantages provided by test marketing. Under order to gauge a product's sales success, it first offers a chance to test it in normal market circumstances. This information not only helps upper management accurately estimate the product's potential national turnover, but it also frequently serves as the foundation for the choice of whether to expand the product nationwide [4]. Therefore, it goes without saying that accuracy is important.

Furthermore, it allows internal management the chance to discover and address any shortcomings in both the product and the marketing strategy while it is still in a phase of limited distribution. This occurs before making the commitment to a nationwide sales launch, a point at which incorporating product improvements and modifications becomes not only challenging but also prohibitively costly [5].

There are three types of test marketing. They all have different objectives, strategies, and intentions. Each has a different set of advantages and disadvantages.

Standard test marketing: Standard test marketing functions similar to a national marketing effort, with the exception of concentrating on a smaller population. It's similar to beta testing in that the advertisement is only seen by a select group of clients. The marketer manages every task involved in a thorough marketing campaign [6]. The primary advantage of standard test marketing is that it evaluates an advertisement's success in the actual world. Standard market testing has one major drawback: other marketers will find out about the new approach or offering [7]. The competitors are given the chance to devise a counter move. Standard market testing also costs money and takes a lot of time. It implies that small enterprises with tight funds should not use this strategy.

Controlled test marketing: Controlled test marketing is a simpler and more affordable substitute for traditional test marketing. New product launches are associated with controlled market trials [6]. The marketer can evaluate the product in many retail locations by putting in place a controlled test market. Suppliers assess checkout scanner data to measure both first and repeat purchases in addition to keeping an eye on competing items' retail sales. These data offer information on how well the new product performs in comparison to its rivals. The primary benefit of controlled test marketing lies in ensuring the distribution of the new product. However, it comes with drawbacks, such as the inability to measure retailer responses and the potential for competitors to gain access to the new product before its official launch. Moreover, the sample size of the test group may not precisely represent the opinions of typical users.

Simulated test marketing (STM): Simulated test marketing strives to provide a more advanced research technique compared to traditional market tests. Simulated market tests are not executed in actual marketplaces [6]. Rather than being conducted in real-world settings, simulated test marketing takes the form of a laboratory experiment. It involves creating a staged marketplace where researchers expose participants to advertising and various variables, aiming to measure levels of purchase intent [8]. Compared to normal market tests, simulated test markets are quicker and less expensive [9]. It enables marketers to concentrate on a specific aspect of their campaign without implementing the entire marketing plan. Figure 13.1 shows the learning from conducting a simulated test market.

Behavioural metrics: Businesses can measure the real actions of participants through STM, including their usage trends, engagement patterns, and purchase decisions. Businesses can gain insights from these behavioural analytics including adoption rates, usage frequency, and possible usage barriers.

Attitudinal metrics: STM facilitates the collection of participant attitudes, viewpoints, and impressions regarding the good or service. Attitude metrics can be gathered by questionnaires, interviews, or feedback forms. Examples of these metrics are satisfaction levels, perceived value, and brand perception. Businesses can assess if a product is market-fit and identify areas for improvement by using these indicators, which offer insights into consumer mood and preferences.

Understanding the "Why": STM makes it possible to go farther and comprehend the fundamental causes of the attitudes and behaviours of consumers. Qualitative research techniques, including in-depth interviews (IDIs) and focus groups, can be used to investigate needs, pain spots, and decision-making processes. With the aid of this qualitative understanding, organisations are better equipped to address their demands by gaining insightful knowledge about the "why."

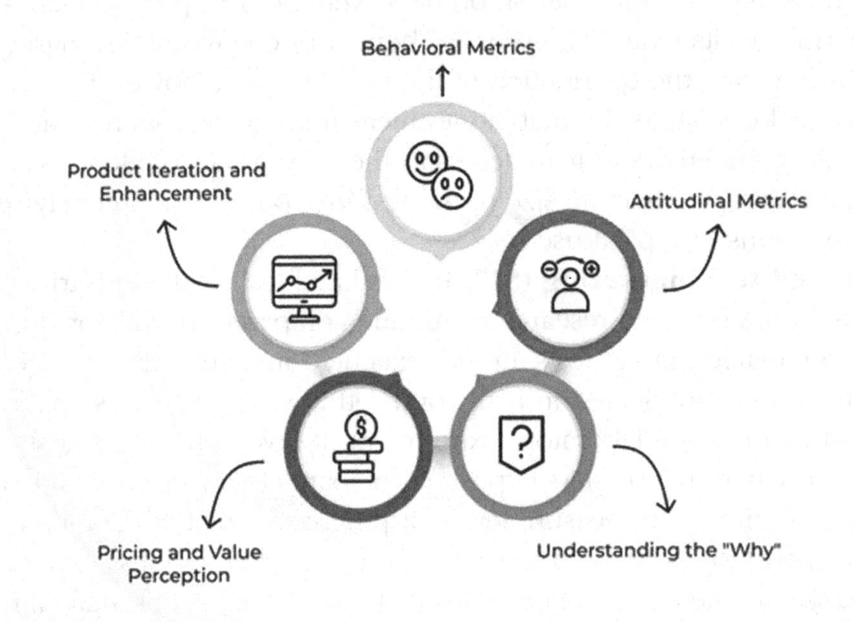

Figure 13.1 Learnings from simulated test marketing [6].

Pricing and value perception: STM offers the chance to evaluate participant views of value and test various pricing strategies. Businesses can evaluate price elasticity, find price thresholds that resonate with the target market, and determine appropriate pricing strategies by monitoring participant reactions to various price points.

Product iteration and enhancement: Prior to the actual launch, STM provides information for product iteration and refinement. This iterative strategy tightly aligns the product with consumer expectations and preferences, increasing the probability of a successful launch.

A typical market test will involve indirect and direct costs [10]: a product manufacturing pilot plant; the production of limited-edition point-of-sale materials; the involvement of an advertising agency; the execution of couponing and sampling initiatives; the allocation of higher trade allowances to secure distribution; and the execution of lower-volume media costs. For example, the cost of a typical two-city test market was $250,000 in 1975 [10]. Four tests were required in order to analyse the two ad weight levels. This translated into a market testing cost of $500,000.

Indirect expenses do exist, though. For instance, there's the expense of disclosing a new product idea to a rival company. The competitor could get to the market if they are more prepared for the launch of new products. Although the price of coming in second in the market is difficult to calculate, marketing expertise suggests that it entails paying more for promotions and a smaller market share. Given the high direct and indirect costs involved in testing the market, the marketing executive should wait to make the decision to proceed until he is certain he has obtained the three essential elements of the ideal product development plan: a successful product, a competitive marketing strategy, and an outstanding communication plan [10].

13.2 Probable Time to Conduct a Test Market

In the new product sequence, opportunities are identified as requirements that are not being sufficiently satisfied by the goods that are currently on the market. This stage of the study often makes use of social psychology-inspired qualitative methods. Group depth interviews or focus groups are often employed techniques. Eight to twelve people will be asked to share their experiences with current items or to recollect how they address particular issues during this two-hour conversation.

A pharmaceutical corporation might, for instance, ask some young mothers how they handle a child's cold. During these conversations, mothers' opinions regarding the advantages and disadvantages of the current items will become apparent. These interviews yield ideas for brand-new items or for enhancing already-existing ones. In reality, these ideas are hypotheses, which can then be investigated through the use of more quantitative techniques like surveys. The procedure of getting rid of all but a handful of the concepts is inherently subjective. It could be predicated on the political climate within the organisation, the expertise of marketing executives, or the executives' own assessment of their industry and capabilities. Capacity should be defined to include executives' ability to function in the industry of the proposed product in addition to production and financial capabilities.

The testing procedures for a product are partly dictated by its anticipated availability. A minimum number of facilities might be needed, for example, to manufacture real goods in order to test a product extension, like adding lemon flavour to an already-existing brand. On the other hand, a completely new production facility is needed for a new home entertainment centre. In the latter instance, the product is tested in the second stage of product testing after the concept has been pretested in the first. The profit plan can then be updated to reflect findings from the study when these concept and product tests are completed. The levels of production costs and, consequently, product acceptance are suggested by these new facts. Extended product use testing may be conducted on a proposed product if it fulfils the minimum payout period.

For a new product, marketing strategy necessitates goal-setting, price, and distribution methods. Product testing may include approximations of price ranges. In the event that your business intends to continue using its current distribution channels, your channel strategy can involve figuring out how to get them to accept the new product. Estimating the costs of an approach like developing a new channel may need a significant amount of work if it is part of the strategy. The creation of the package, the choice of the intended media mix, and the copy theme for commercials are all included in the communication strategy.

It is significant to remember that the creation of the marketing mix plans for strategy development is influenced by the communication plan. The ultimate estimated profit plan uses the communication plan's anticipated cost as input [10]. Next, in order to move forward to the test market stage, the proposal is presented to the management for approval. Hence, an entrepreneur can test the product only after they have assumed that they have a successful healthcare product or service.

The final phase of a procedure that aims to create a successful product is test marketing [4]. The "ideal" process starts with the ideation of new products and concludes with a thorough marketing plan review. As we show in the exhibit, the initial step of this process is intended to produce a large number of ideas with little to no review. After that, these concepts go through a number of screenings based on standards set by company policy [11]. For example, the exhibit's first screen is financial in nature, showing an approximate profit projection that contrasts the projected payout term with the longest duration permitted by company regulation. As ideas go through subsequent screens in the development of products and communications, they are evaluated once more. If an idea doesn't make it past a screening, it might be abandoned or given another look and revision.

The outcomes of the advancements in product and communication should be combined with the marketing strategy to build a comprehensive marketing plan. From this point on, the process needs to focus more on fine-tuning and plan evaluation than on creation or development. If concept development and idea generation occur during the test market phase, the management's reluctance to discontinue items sooner is evident. Early in the screening process, a low "kill ratio" suggests that either the management is circumventing the criteria or it is not strict enough.

When a product concept looks nice, product managers usually say "go," whereas researchers prefer to say "stop" until they see something better. This philosophical gap arises from the different ways that researchers and product managers are viewed: researchers are supposed to save expensive mistakes, whereas product managers are rewarded for successful products. When establishing the criteria for evaluation in the new product development process, these philosophical differences clash. Higher dosages for earlier kill are preferred by researchers. Lower levels would be preferred by product managers in order to test-market more goods. By creating screening standards that align with company goals, available resources, and available opportunities, the marketing executive must make these disparities right [11].

Marketing executives have to set these standards as well as the procedures for overriding them. Scholars concur that overrides are necessary, but they contend that an override ought to be recognised as a system exception [12]. Without limiting the product manager's creativity, this notation shields the researcher from unjust judgement. An executive finds a test market more trustworthy than any other test procedure because of its actuality and integrative nature. This helps to explain why CEOs tend to ignore other data in favour of testing market findings. A secondary rationale for only depending on test markets is the challenge of assimilating the multitude of data and assessments at the marketing executive's disposal. However, the marketing executive can get assistance from freshly created simulation models in putting information together to decide on new products.

There are also a few recent developments in the field of test marketing as follows:

Laboratory simulation: In this procedure, test advertisements are shown to a sample of customers before allowing them to make purchases in a fictitious supermarket. Data about brand preferences, the economy, and demographics may be obtained before the test. After home use, follow-up interviews are conducted to gauge customer satisfaction and repurchase intentions. Subsequently, these data might be included in mathematical models to forecast market share [10]. Following the lab simulation, there are three potential options. First, the product might be completely discontinued. Positive outcomes from the laboratory simulation could also persuade the marketer to proceed straight to a regional launch. Third, a test market might be held after the lab simulation to gauge the marketing plan's effectiveness, including trade acceptance, make any adjustments, and act as a safety net [1].

Mathematical market testing: Via order to anticipate market share in mathematical simulations, marketing experts have created correlations between promotional expenditures, awareness, trial, and repeat purchase. Forecasts of market share are used as inputs in the profit plan to determine the payout period. First, a preliminary payout estimate is developed using simulation models. Executive judgements serve as the model's inputs at this point in the procedure. In the subsequent payout period estimate, executive judgement is replaced with product-test data. After that, the estimate might be updated once again by including data from the communication and extended product-use tests into the simulation model [10].

In summary, the entrepreneur ought to save the decision to test the market until after obtaining a detailed marketing plan that is predicated on exhaustive product and communication testing. Test markets should be used to evaluate how well the marketing strategy is working in terms of trade and customer reaction, find methods to increase the productivity of the plan, and prevent any potential problems. The test market's outcomes ought to be utilised to assess the marketing strategy rather than to come up with new concepts or judge individuals [13].

13.3 Factors Which Help in Decision Making

Testing the market should never be a routine decision. Test markets should only be utilised as a last resort because gathering data on consumer reactions to new items requires a lot of money and work [14]. If a new product's risk is manageable and its research is sufficiently encouraging, it is more profitable to introduce it nationally rather than through a test market, which comes with expenses and delays. The corporation can reduce losses through test marketing, but not increase profits. It's critical to strike a balance between a product's cost and failure risk and its potential for profit and success. For instance, in the last three years, Cadbury Typhoo Limited has test-marketed 24 goods; yet, in the same time, has also successfully released four products across the UK without using a test market phase.

Whether or not to test depends in large part on how different the test's financial requirements are from those of the national launch route [15]. For the items we picked for a test launch vs those we have immediately placed onto the national market, very little change in manufacturing investment was needed [16]. On the one hand, the test launch strategy is more advantageous when the plant investment is significant for a national launch but negligible for a test market. However, there might be a significant opportunity cost if a product is limited to the test market area for the length of time needed to accurately anticipate performance. Depending on how long the product is kept in test, this opportunity cost could equal the turnover and profit that would have been made from one year's worth of national sales [17].

If the test is successful, another thing to take into account is how quickly and likely it is for the competition to imitate your product and take over a portion of your domestic market or international markets. Rivals will keep an eye on the test market and, if equipped with the necessary technology, will create their own iterations of your product. Two years after Cadbury launched their children's chocolate brand, Curly Wurly, in the United Kingdom with great success, rival products bearing identical names started to show up in West German, Canadian, Japanese, and American markets.

All new product launches come with a significant marketing budget, which varies depending on the size of the launch, in addition to any potential investments in plant and machinery. When a new product is introduced, the company must invest heavily in advertising and promotion; the sales force must devote time, attention, and effort to the new product; and shelf space in wholesale and retail establishments must be secured, sometimes at the expense of space already occupied by the company's current offerings.

Furthermore, in the event that a new product fails, expenses related to writing off undesired and unusable materials and packaging must be incurred in addition to the costs of rebating and recovering unsold stocks from customers. Top management should also consider the potential harm that a new product failure could do to the business: it could damage the company's reputation among customers, which is a real, albeit unquantifiable, risk.

References

1. Lim WM, Ting DH. Healthcare marketing: contemporary salient issues and future research directions. *Int J Healthc Manag* 2012;5:3–11. https://doi.org/10.1179/204797012X13293146890048.
2. Buchta C, Dolničar S. Learning by simulation -computer simulations for strategic marketing decision support in tourism. *Int J Tour Sci* 2003;3:65–78. https://doi.org/10.1080/15980634.2003.11434540.
3. Bierbooms JJPA, Bongers IMB, Oers HAM van. Strategic market orientation in mental healthcare: A knowledge synthesis. *Int J Healthc Manag* 2012;5:141–53. https://doi.org/10.1179/2047971912Y.0000000012.

4. Ben Ayed M, El Aoud N. The patient empowerment: a promising concept in healthcare marketing. *Int J Healthc Manag* 2017;10:42–8. https://doi.org/10.1080/20479700.2016.1268326.
5. Fortenberry JL, McGoldrick PJ. Internal marketing: a pathway for healthcare facilities to improve the patient experience. *Int J Healthc Manag* 2016;9:28–33. https://doi.org/10.1179/2047971915Y.0000000014.
6. Demystifying simulated test markets: goals & objectives n.d. https://blog.market-xcel.com/what-is-stm-goals-and-objectives-of-simulated-test-market/ (accessed January 4, 2024).
7. Three types of test marketing and their importance - INK n.d. https://inkforall.com/ai-writing-tools/copy-testing/three-types-of-test-marketing/ (accessed January 4, 2024).
8. Altsitsiadis E. Marketing health care simulation modelling: towards an integrated service approach. *J Simul* 2011;5:123–31. https://doi.org/10.1057/JOS.2010.12.
9. Woodham OP. Testing the effectiveness of a marketing simulation to improve course performance. *Mark Educ Rev* 2018;28:203–16. https://doi.org/10.1080/10528008.2017.1369356.
10. Test marketing in new product development n.d. https://hbr.org/1976/05/test-marketing-in-new-product-development (accessed January 4, 2024).
11. When, where, and how to test market n.d. https://hbr.org/1975/05/when-where-and-how-to-test-market (accessed January 4, 2024).
12. Cueny D, Miller K, Eldridge MK. The healthcare account executive-a sales approach to healthcare marketing. *Health Mark Q* 1985;3:85–92. https://doi.org/10.1300/J026V03N02_11.
13. Hunt SD. Advancing marketing strategy in the marketing discipline and beyond: from promise, to neglect, to prominence, to fragment (to promise?). *J Mark Manag* 2018;34:16–51. https://doi.org/10.1080/0267257X.2017.1326973.
14. Walker D, Knox S. Understanding consumer decision making in grocery markets: does attitude-behaviour consistency vary with involvement? *J Mark Commun* 1997;3:33–49. https://doi.org/10.1080/135272697346032.
15. Valos MJ, Maplestone VL, Polonsky MJ, Ewing M. Integrating social media within an integrated marketing communication decision-making framework. *J Mark Manag* 2017;33:1522–58. https://doi.org/10.1080/0267257X.2017.1410211.
16. Goslar MD. Capability criteria for marketing decision support systems. *J Manag Inf Syst* 1986;3:81–95. https://doi.org/10.1080/07421222.1986.11517756.
17. Pitf LF, Nel D. The effects of group cohesiveness on decision performance in a simulated marketing environment. *South African J Sociol* 1990;21:59–65. https://doi.org/10.1080/02580144.1990.10432111.

Chapter 14

Mechanical Testing and Characterisation of Products

14.1 Mechanical Testing

Development of a healthcare product requires several techniques of manufacturing be it computer numerical control (CNC) machining, injection moulding, 3D printing, etc. In order to test whether the product will last in actual usage scenarios, mechanical testing is required. Mechanical testing, in general, is mandatory in determining the overall strength of the product to ensure its durability and quality. It can also be used to ascertain the material selection for a certain product.

One method of evaluating and quantifying a material's qualities is mechanical testing [1–3]. Compression, flexure, tension, and other tests can be performed on materials or completed products to measure the forces that may cause the material to break, snap, bend, or yield during the test. It is crucial to conduct mechanical testing to evaluate a material's characteristics or validate finished goods to make sure they are safe and suitable for usage. In order to build a product with the required functionality, the mechanical designer would want to select the components with the necessary qualities during the design phase [4]. After a product is designed, the designer must make sure that it can be manufactured repeatedly, consistently yielding the required results, and safely for the intended user base.

The goal of fundamental mechanical testing is to ascertain the mechanical characteristics of a single material, regardless of its geometry. Because batches of raw materials can vary, mechanical testing of each batch guarantees that it

DOI: 10.4324/9781003475309-19

satisfies minimum strength standards. In order to measure a composite material of a specific size or to ascertain how a structure or final product will react to a particular action—for example, how an implant will respond to cyclic loading or whether artificial tissue that has been developed will resemble the original—empirical or imitative testing is also available.

To further proceed with the chapter, some terminology should be understood. This terminology is regularly used in mechanical engineering to determine the properties of a material.

- Elastic and plastic deformation: Deformation, or a change in an object's size and shape, is essentially the same as strain. It happens as a result of a force being applied or a temperature change. Various types of deformation can occur based on factors such as material composition, size, and applied force. First of all, once the applied forces are removed, elastic deformation is reversible and goes away. For example, stretching a rubber band. When an object experiences non-reversible changes in size or shape due to applied force, this is known as plastic deformation [5]. The shape doesn't change even after the force is removed. Consider the way steel rocks bend.
- Stress: Stress is a physical variable in continuum mechanics that characterises forces experienced during deformation. A stretched elastic band, for instance, is sensitive to tensile stress and may elongate when it is pulled apart. Compressive stress can cause an object to shorten when it is forced together, like a crushed sponge.
- Strain: Elongation (or strain) is the ratio of an object's deformation to its original length, while deformation is a measure of how much an object is stretched [6]. Consider strain as the stretch percentage, or the amount that an object expands or contracts under load.
- Yield strength: The yielding capability or malleability of an object is determined by its yield strength. It's the moment when something stops being elastic and turns into a plastic substance. Yield strength aids in the selection of suitable materials based on specifications. For instance, metal would not have been able to be moulded into the unusual forms, so plastic was used instead of metal while manufacturing hearing aids.

Other terminology related to different test methods will be explained in their respective sections. In the rest of the chapter, several common and recent advances in mechanical testing methods are discussed to give an overview of the testing procedures and industry.

14.1.1 Tensile and Compressive Test

There are various reasons why tensile and compression tests need to be carried out. Tensile test results are considered when choosing materials for technical applications. To guarantee quality, tensile properties are usually mentioned in material specifications. Tensile properties are frequently investigated when creating new materials and procedures so that various materials and procedures can be contrasted. Ultimately, a material's response to loading scenarios other than uniaxial loading is frequently predicted using its tensile or compressive properties [3, 7]. In this method, a material is generally stretched or compressed using grippers (Figure 14.1) and the material behaviour is recorded. These tests are usually performed on a universal testing machine (UTM) (Figure 14.2). UTMs also come in several sizes depending on the application.

An operator must carry out a number of duties to guarantee that a tensile test is carried out in compliance with internal and/or external testing standards. These processes may be fully or partially automated, depending on the lab; nevertheless, the operator is always ultimately responsible for ensuring that the setup is proper. To start with, firstly a specimen or a sample is manufactured using the same material which is to be tested. This specimen's design is as per the international standards (i.e., American Society for Testing and Materials (ASTM) and International Organization for Standardization (ISO)).

Specimen geometries might vary greatly depending on the substance being tested, the test protocol, and the standard being adhered to. Organisations that generate standards, such as ASTM and ISO, have created specimen standards for a range of materials, making it possible to compare the characteristics of different batches and manufacturers with confidence.

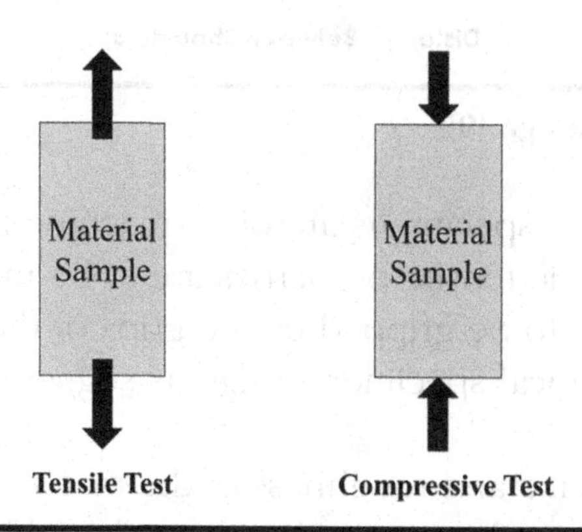

Figure 14.1 Schematic for tensile and compressive tests.

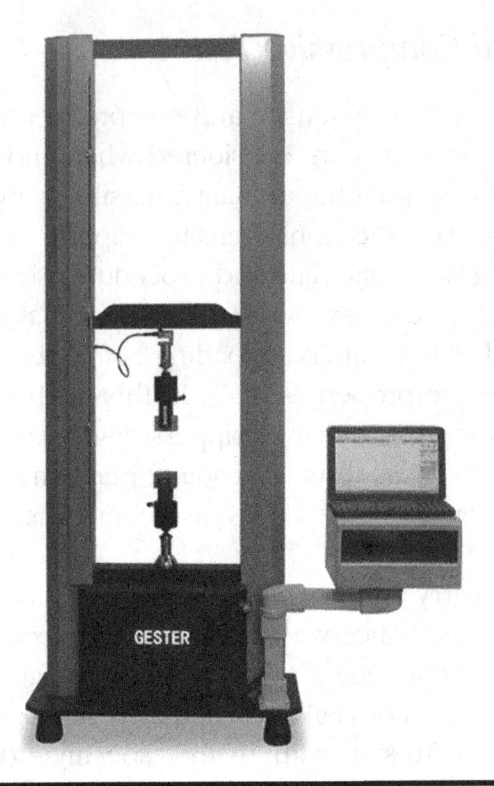

Figure 14.2 A typical universal testing machine (UTM) [8].

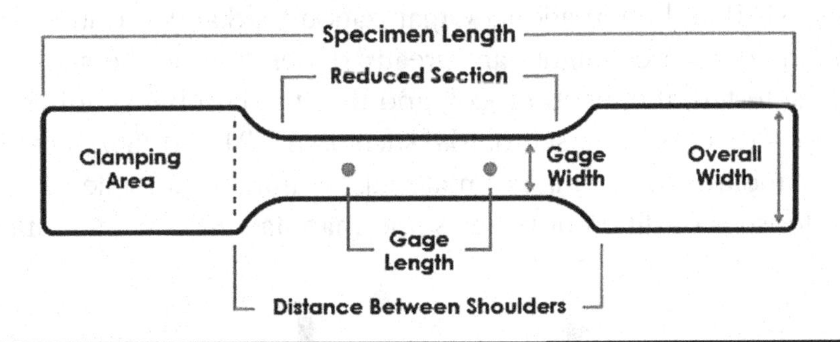

Figure 14.3 Specimen design [9].

Dogbone-shaped tensile specimens are often machined or die-cast, with a "gauge length" where the tensile properties are to be measured and "shoulders" that are intended to be gripped by the grips of the testing apparatus. Figure 14.3 shows a typical specimen design as suggested by the internal standard committees.

Depending on the size and roughness of the material, many grip types and jaw face surfaces may be required to properly hold the specimens. Grips come in a wide variety of force capabilities and surface kinds, including

Figure 14.4 Insertion of specimen into the machine [9].

rubber-coated, serrated, smooth, and others. To assist operators with aligning a specimen into the grips and applying force in the proper direction, a variety of alignment devices are available. Figure 14.4 shows the insertion of the specimen into the device.

It's time to begin the test after the specimen has been placed into the apparatus and the extensometer has been attached. When configuring the test, it is required to input the right test method in the testing programme and provide parameters such specimen measurements, test speed, and end criteria. The apparatus will apply tensile force to the specimen in accordance with the test protocol after the operator gives the go-ahead, recording data as the specimen reacts to the stress. The specimen can be taken out once the test is finished, and the data can be exported for additional research.

Figure 14.5 shows a typical graph generated by the UTM during a tensile testing of a steel sample. This graph is generally known as the stress-strain graph. As the name suggests, this graph compared the stress-strain behaviour of the specimen during the stretching until its breakage (or fracture). The tensile test is a test that can be used to determine a material's strength. The content is ruthlessly extracted from both ends in this test. The stress-strain graph represents the relationship that may be graphed between the strain an object experiences as a result of the stress it is subjected to.

A material that is subjected to excessive stress will become permanently deformed, and when the stress is applied, a "neck" will form in addition to the distortion. The neck will snap under much more tension. Eventually, the material breaks due to the lack of stress. The graph shows certain terminology, which is explained below.

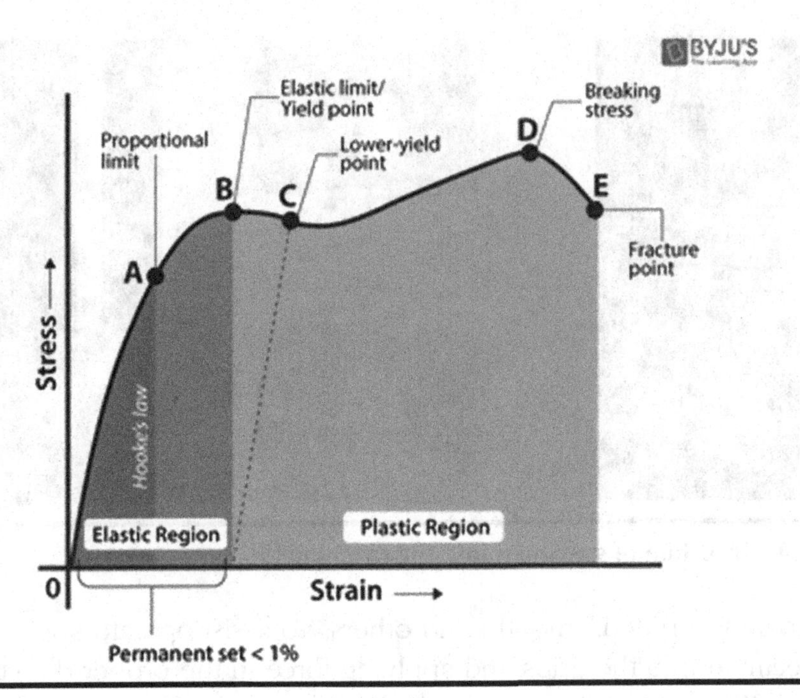

Figure 14.5 Stress-strain of a steel sample generated by a universal testing machine [10].

Proportional limit: Hooke's Law governs the area of the stress-strain curve. The stress-to-strain ratio in this limit provides us with a proportionality constant called Young's modulus. On the graph, the point OA represents this concept.

Elastic limit: This is the diagrammatic point at which no loading occurs and the material reverts to its initial state. After this, the material no longer regains its original shape and starts to exhibit plastic deformation.

Fracture point: It indicates the point on the stress-strain curve at which the material breaks down.

14.1.2 Torsion Test

Test components or materials are subjected to torsion tests, which twist them to a predetermined degree, with a predetermined force, or until the material fails [11]. Torsion tests involve attaching one end of the test sample to prevent it from moving or rotating and providing a moment to the other end to cause the sample to revolve around its axis [12, 13]. This process applies the twisting force to the test sample. The sample's two ends may likewise be subjected to the rotating moment, but they must be spun in opposing directions. Figure 14.6 shows a typical schematic of torsion tests performed on a shaft, where, L is the shaft length, D is the diameter and Θ is the torsion angle.

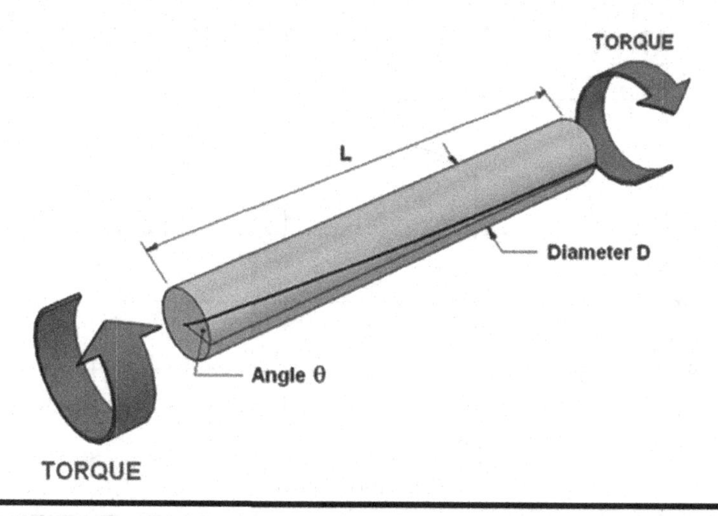

Figure 14.6 Schematic of load application during a torsion test over a specimen [14].

Three types of torsion testing are frequently carried out: operational, proof, and failure. The sample's strength is determined by twisting it until it breaks in order to test for failure by torsion [15]. The purpose of a proof test is to monitor a material under a given torque force for a predetermined amount of time. Lastly, operational testing assesses how well a material performs under the anticipated service circumstances of its intended use. Any of these test types can be performed with either torsion-only loading or a combination of torsion and axial (tension or compression) loading, depending on the properties that need to be evaluated. Figure 14.7 shows the graph obtained by a torsion testing machine. It usually generates a torque vs angle graph where the twist resistance of a material can be judged. The vertical line at the end shows the failure point where the material finally breaks due to twisting. In this case, the maximum angle was observed to be 60 degrees at approximately 0.5 N-m torque.

Torsion testing covers more than just raw materials; it also includes final goods. Torsional stresses are a common occurrence in the daily functioning of biomedical tubing, switches, fasteners, and several other equipment. Manufacturers examine product quality, confirm designs, replicate real-world service circumstances, and guarantee the suitability of production processes by putting these goods through torsion tests. Materials used in the structural, automotive, and biomedical industries have additional applications requiring torsion [17]. These materials can be constructed of metals, polymers, wood, or ceramics, among other materials, and are frequently shaped like bolts, rods, beams, tubes, and wires. Figure 14.8 shows the different applications for torsional tests. In the first part, a plastic container is tested. In the second part of the figure, a key and lock is tested, whereas in the last part, a part of a nebuliser is tested.

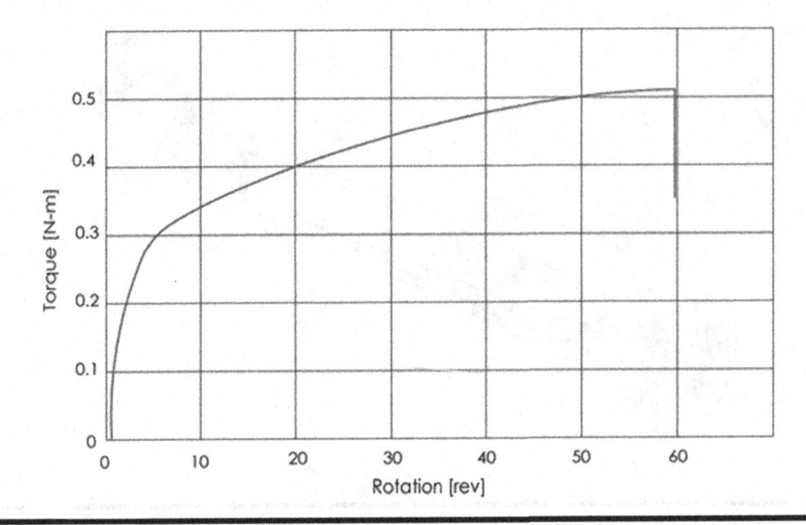

Figure 14.7 Typical graph obtained from a torsion test [16].

Figure 14.8 Typical applications of torsional testing [16].

14.1.3 Fatigue Test

The results of a fatigue test are used to assess a material's resistance to cyclic fatigue loading scenarios [18]. Cycle fatigue tests involve repeated loading and unloading in tension, compression, bending, torsion, or combinations of these stresses. Common loading conditions for fatigue tests include tension into tension, compression into compression, and vice versa.

Failure is typically the result of cyclic loading, regardless of the stress level. The S-N curve (Figure 14.9) does, however, level off for some materials, indicating that for these materials a fatigue limit—a defined stress or

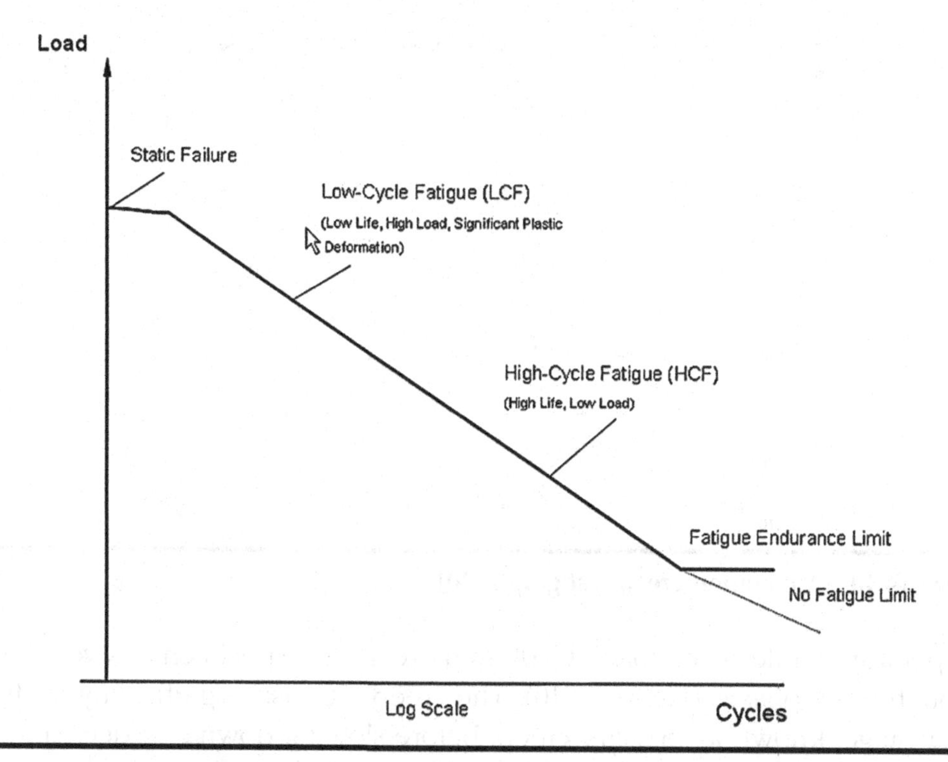

Figure 14.9 A common fatigue cycle graph, also known as S-N curve [19].

load—may be established below which an unlimited lifespan can be anticipated. In Figure 14.9, if a product or specimen fails at low cycles of load, it is termed as static failure. Whereas, when a material endures high number of cycles, it is rated as high-cycle fatigue.

In order to conduct a fatigue test, a sample is deflated to either zero load or the opposite load after being placed inside a fatigue tester or machine that is powered by the predefined test voltage. Until the test is finished, this cycle of charging and discharging continues. Depending on the requirements, the test may be run for a predetermined number of cycles or until the sample fails. In addition to the two main types—load controlled high cycle and strain controlled low cycle—there are other common methods of fatigue testing.

14.1.4 Creep Test

A creep test, commonly called a stress-relaxation test, examines the amount of deformation a material experiences over time when exposed to a continuous tensile or compressive load at a constant temperature. Creep tests are necessary for materials that must be able to maintain a particular operating

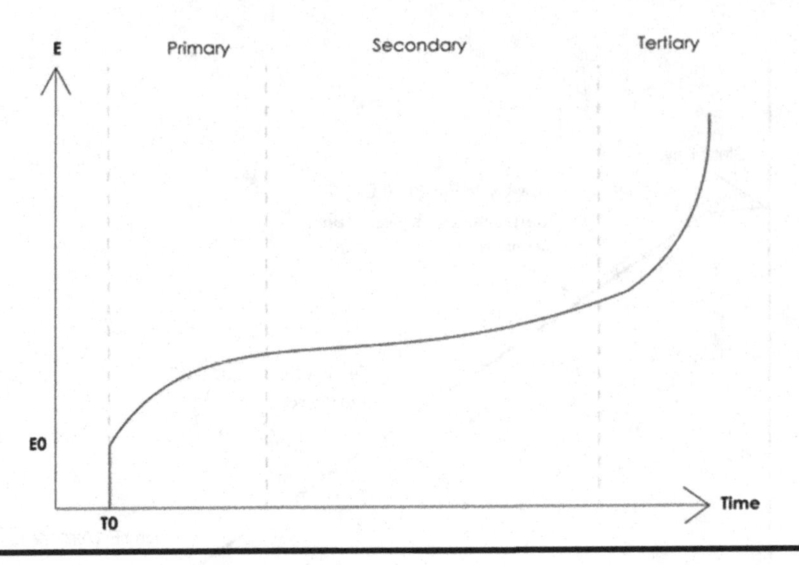

Figure 14.10 A common creep test graph [20].

temperature while being loaded [20]. In a creep test, specimens move through three phases (Figure 14.10). The creep rate rises significantly in the initial stage, known as primary creep, before slowing down and decreasing. The secondary stage is characterised by a rather constant creep rate, whereas the tertiary stage is the creep stage.

14.1.5 Hardness Test

The resistance of a substance to deformation at its surface is known as its hardness. The resulting indentation is measured and translated to a hardness value based on the test's specific hardness scale. There are numerous standardised hardness tests available, and each one can be used with a variety of materials, including rubber, metals, and ceramics. Comparing materials and ensuring quality control in the production and hardening processes are two advantages of measuring hardness [21]. Hardness is a measurable attribute rather than a fundamental physical property of a substance. However, depending on the intended use, it can offer some useful information regarding a material's strength and durability.

To measure the hardness of a material, a hard, standardised instrument is pressed into the material to be assessed. An indenter is a tool that is pushed with a preset force into a sample for a specific period of time (measured in seconds). There is a quantifiable amount of surface deformation when the indenter is pressed into the material. Following that, the deformation is measured, and the measurement is used to assign a hardness value to the material according to the specific scale of the test.

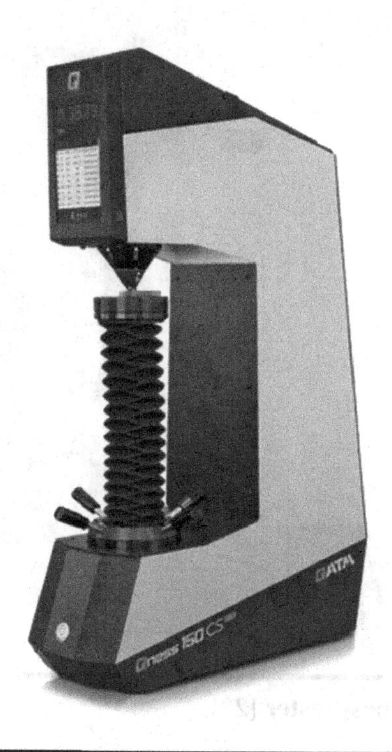

Figure 14.11 Rockwell hardness tester [22].

Testing for hardness is crucial for several reasons. One of the most important factors in a material's application can be its hardness. This can apply to metal parts that are cut and ground, or, on the other extreme, to rubbers that are used to absorb shock. The ability to test and evaluate material hardness objectively is crucial. A variety of hardness test techniques exist, each appropriate for a particular material type or measurement scale. As a result, depending on the chemical being examined, various procedures should be chosen. The following describes the most popular hardness tests:

Rockwell hardness: It is intended to give a quick readout, especially for metal samples. The measurement of the depth of the deformation generated by the indenter determines the number of the indentation; a lower number indicates a softer material. The Rockwell regular hardness test and the Rockwell superficial hardness test are the two main test types employed. Figure 14.11 shows a typical Rockwell hardness tester.

Brinell hardness: The hardness value is obtained by measuring the diameter of the indentation made by a spherical indenter. Given that it produces a larger indentation that is less susceptible to local differences in material hardness, it is typically employed for evaluating larger samples or samples that are not entirely homogeneous.

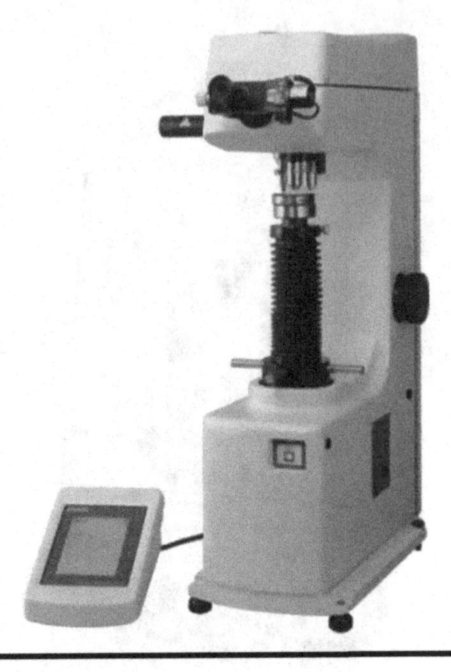

Figure 14.12 Vickers hardness tester [23].

Vickers hardness: The test utilises a diamond pyramid to create a square impression on the material's surface for testing. The hardness value is obtained by converting the optical measurement of the indentation's diagonal. Micro-testing can be done using the Vickers hardness test. Figure 14.12 shows a common Vickers hardness machine.

Knoop hardness: This is used especially to evaluate brittle or thin materials because it provides a useful reading even for shallow indentions. It creates the impression with an elongated pyramid, and the Knoop hardness is determined by measuring the length of the diagonal. Testing of micro-hardness is done with it.

Shore hardness: This is applied to softer materials like plastics and elastomers. The material sample is forced into a durometer, a spring-loaded indenter that converts the depth of penetration to a hardness measurement, scales ranging from Shore A to Shore D based on the hardness level. Figure 14.13 shows a durometer which is used to measure shore hardness.

14.1.6 Tribotesting

A device that simulates wear and friction at the interface between surfaces in a relative motion under controlled conditions is called a tribotester or tribometer. Tribometers are created and employed for many different uses,

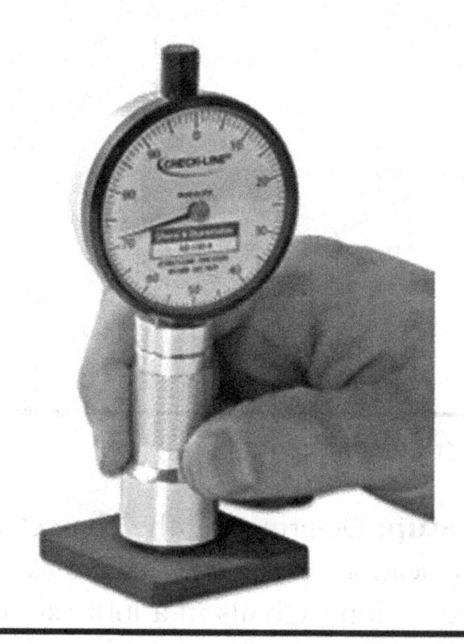

Figure 14.13 Durometer [24].

such as but not restricted to mimicking the tribo-contact conditions found in a specific machine, analysing potential bearing materials for an application where friction is crucial, assessing lubricants in light of a specific application, tracking surface contaminants on a product and gathering generic wear properties of materials, and examining the basic principles underlying solid or lubricated solid friction. The choice of a certain tribometer is guided by the objective of the test.

Test design needs to take into account the main factors influencing friction [25]. Numerous tribometers are in use to replicate a variety of scenarios found in the real-world applications described below, as tribological issues arise in nearly every branch of engineering [26]. Even though commercial tribometers are widely accessible, researchers are still developing customised tribometers. This is required to supply specialised component simulations, testing settings, or specimen dimensions that are not easily accessible. There are several types of tribometers available in the industry according to the application and type of materials.

Pin-on-disc: In tribology, pin-on-disc tribometers are arguably the most well-known and often utilised instruments. A revolving disc and a stationary pin make up the tribometer. A dead weight or actively managed systems load the pin. Pins come in three main shapes: spherical, triangular, and flat. The latter situation is known as the "ball-on-disc" test. Figure 14.14 illustrates the pin on disc (really ball-on-disc) setup schematically.

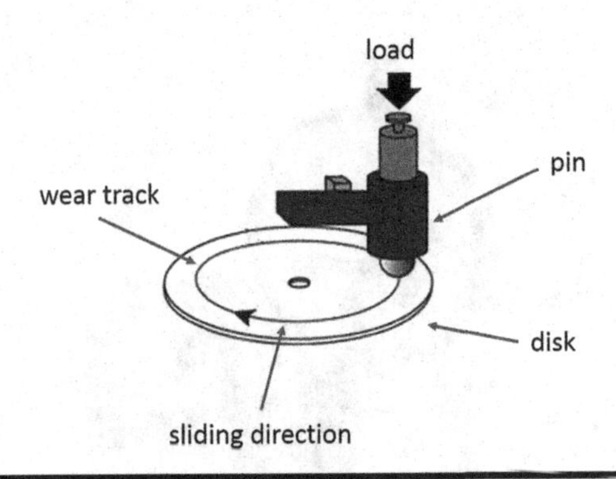

Figure 14.14 Pin-on-disc tribometer [25].

Four-ball testing setup: Determining the characteristics of greases and lubricants for sliding applications is the test's main goal. A ball is rotated under load against three stationary balls in a lubricated environment as part of the test. Standards have been set, and measurements are made at various speeds, temperatures, and times. Lubricants can be compared using the scar sizes found from wear tests. After the examination, the scar's size is evaluated to see how well the lubricant stops wear. Because of the applied loads, each ball in the wear test has a circular wear scar; for comparison, the average wear scar diameter is computed. The steel balls' wear scar diameter is measured using an image capture system.

Wear test and reciprocating sliding friction: The average coefficient of friction (COF) and wear may be determined using the Reciprocating Sliding Friction and Wear test. Tests can be performed in either a lubricated or dry environment. Contact geometries can take several forms, including ball-on-disc, ball-on-plate, cylinder-on-disc, cylinder-on-plate, and disc-on-disc.

Fretting tester: Fretting is a kind of wear described by low amplitude rhythmic sliding between bodies that are nominally at rest (for example due to vibration, cyclic stresses, etc). In bolted joints and electrical connections, the sliding amplitude can range from tens of nanometres to tens of microns. Thus, fretting tribometers are specialised instruments designed for small-amplitude reciprocating motion.

14.2 Non-Destructive Testing

Non-destructive testing (NDT), alternatively referred to as non-destructive inspection (NDI) and non-destructive evaluation (NDE), is a multidisciplinary field that merges quality assurance with materials science. The purpose of NDT is to assess and examine materials, assemblies, and parts without jeopardising their usability [27]. Skilled technicians utilise a range of testing methods to detect inclusions, voids, cracks, and weld discontinuities. They also identify incorrectly fabricated subcomponents. NDT is essential for overseeing industrial processes, reducing production costs, ensuring product integrity and reliability, and upholding a consistent quality standard. The absence of NDT poses substantial risks to the reliability and safety of the components [28]. As a result, NDT plays a crucial role in averting catastrophic failures such as pipeline explosions and leaks, aircraft and locomotive collisions, nuclear reactor malfunctions, and maritime disasters. It encompasses various NDT methods that warrant discussion:

Acoustic emission testing (AE): This kind of NDT focuses on picking up the brief ultrasonic flashes that active cracks under load produce [29]. The AE is detected by sensors positioned across the structure's surface. In severely strained locations, AE can even be identified through plasticisation before a fracture appears.

Liquid penetrant test: Applying a low viscosity solution to the substance to be examined is known as liquid penetrant testing [30]. This fluid seeps into any irregularities like cracks or porosity before a developer is applied, allowing the penetrant liquid to creep upward and generate visual proof of the fault. In liquid penetrant testing, water washable, post-emulsifiable, and solvent-removable penetrants can all be utilised.

Thermal or infrared testing: By using sensors to measure the wavelength of infrared light emitted from an object's surface, infrared testing, also known as thermography, can be used to evaluate the state of an object. With the use of sensors, passive thermography measures the wavelength of radiation emitted. If the emissivity is known or can be inferred, the temperature can be computed and shown as a false colour image or digital measurement. This is frequently used to track heat loss from buildings and is helpful for identifying overheating bearings, motors, or electrical components. A temperature gradient is created inside a structure using active thermography. Its internal features that impact heat flow cause fluctuations in surface temperature, which can be analysed to assess a component's health.

Ultrasonic testing: High frequency sound is transmitted into a material during ultrasonic testing in order to interact with its reflective or attenuating properties. The three main categories of ultrasonic testing are Time of Flight Diffraction (ToFD), Through Transmission (TT), and Pulse Echo (PE).

Visual testing: One of the most popular methods, known as visual testing or visual inspection, is the operator observing the test component. Utilising optical tools like magnifying glasses or computer-assisted systems (sometimes referred to as "Remote Viewing") can help with this. This technique makes it possible to find damage, cracks, corrosion, misalignment, and other issues. The majority of other NDT methods include visual inspection by default since they usually call for an operator to search for flaws.

There are several clear benefits, the most notable of which is that the test components are not harmed during the procedure, meaning that if any issues are discovered, the item can be fixed rather than replaced [28]. Because most of the approaches are safe for humans to use, it is also an extremely safe testing method for operators. Because NDT is reproducible and allows for the correlation of results across multiple tests, it is also a fairly accurate method of inspection. These methods of testing are also cost-effective. NDT is less expensive than destructive testing because it can stop a malfunction before it starts without damaging the component.

14.3 Recent Advances in Mechanical Testing for Healthcare

The above-mentioned testing techniques are widely used in the healthcare sector. In this section, few applications of tensile testing and tribotesting are mentioned with respect to healthcare engineering.

14.3.1 Mechanical Modelling of Brain Tissues to Develop Artificial Surrogates

One of the functional tissues that has been examined the most in recent decades is the human brain. The human brain is still a mysterious part of the human anatomy, despite decades of research [31]. Traumatic brain injury (TBI) is a disorder of the brain tissue that results from mechanical hits, such as car accidents and falls, that can cause temporary or permanent brain damage [32]. It has been discovered that these wounds have a major impact on the mechanical characteristics of brain tissue [33]. Furthermore, variations were observed in the mechanical properties of brain tissue for several disorders, including hypertension, dementia, and Alzheimer's disease.

To mechanically characterise and develop artificial surrogates of brain, Singh et al. [34] developed the surrogates using a multi-part polymeric

Figure 14.15 Test coupon moulds [34].

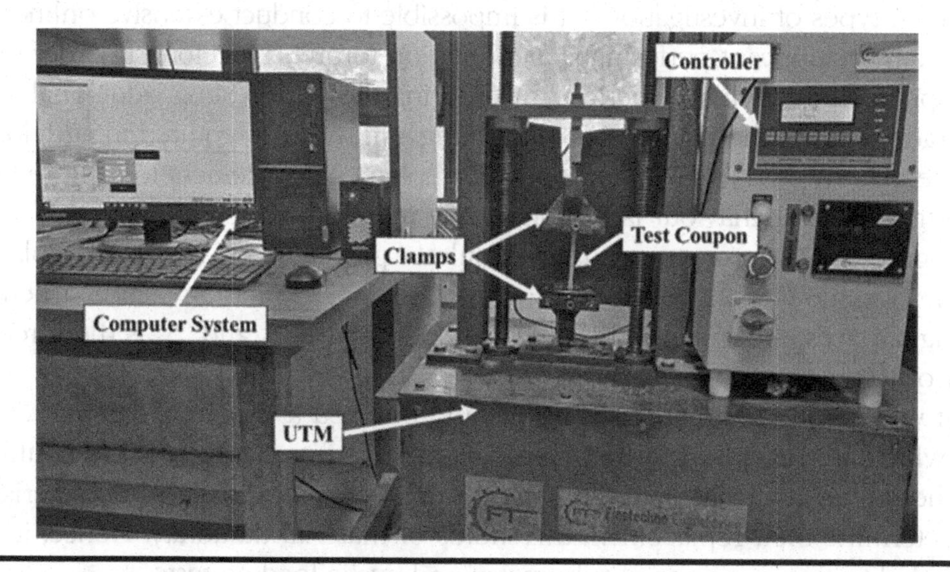

Figure 14.16 Universal testing machine setup [34].

material. Two-part polymeric materials with varying Shore hardness were combined to create test coupons with a range of Shore hardnesses from 5 A to 30 A matching to that of a brain. Variations in the concentration of polymeric material were used to create a total of 15 test coupons for brain tissue surrogates that had the same dimensions (50 mm length, 10 mm breadth, and 3 mm thickness) (Figure 14.15). Furthermore, the coupons were tested through tensile tests (Figure 14.16).

Three compositions were found to closely resemble brain tissue out of the 15 compositions that were analysed. To ensure repeatability, these designated control specimens underwent a thorough testing procedure. Hence, these methods could be used to develop artificial surrogates which match the exact mechanical properties of the concerned organ.

14.3.2 Tribo-Evaluation of Hip Implants

Surgery for total hip replacement (THR) is regarded as an effective treatment for degenerative joint disorders. Hip joint wear testers, also known as hip joint simulators, are used to evaluate how well hip prostheses wear over time in body-simulated settings that mirror the dynamic and kinematic aspects of the human hip joint. [35]. A five-million-cycle hip simulator wear test typically takes more than three months to complete. Furthermore, it is thought that the wear rates acquired from interval monitoring are the most essential outcomes of these types of investigations. It is impossible to conduct extensive online research on lubricant performance and friction torque. Therefore, in order to comprehend the dynamic performance during the wear testing, additional characteristics of the prosthesis's articulating surfaces are required in addition to the material loss [36]. An appropriate supplementary element to capture the real tribological behaviour is frictional force monitoring.

Figure 14.17 shows a typical hip simulator presented by Hua et al. [36]. "1" is the slop module, "2" is the housing for bearing, "3" is the reaction articulating surfaces, "4" is the friction measurement module, and "5" is the force sensor to measure the friction.

It was mentioned that this device could be used to test new materials which are proposed for hip replacement implants. The test setup could further be used to calculate the friction and wear properties of the materials to determine their replacement thresholds or material durability. Hence, it included testing concepts of tribometry and cyclic loading tests.

14.3.3 Testing of Anisotropic Artificial Skins Using Tensile Testing Machine

Anisotropic and heterogeneous soft tissues include the skin, tissues, and organs [37–39]. There has been limited study into anisotropic synthetic skin, and none of these studies have included biaxial testing to establish whether or not the skin can correctly match the mechanical characteristics of various body areas. Gupta et al. [38] created a soft composite material system to imitate anisotropic skin types as part of their research. The influence of

Figure 14.17 Hip simulator [36].

layer count and fibre orientation on the biomechanical properties of skin has received a lot of attention. To properly define the synthetic skin, many hyperelastic material models were utilised in the characterisation procedure. Figure 14.18 shows the biaxial tensile tester used to study the artificial skin samples.

Several components such as, stress-strain behaviour and ultimate tensile stress were computed and compared with the actual skin sample values present in the literature. All things considered, this research was successful in creating an unique biofidelic material system that faithfully mimics the characteristics of human skin, enabling better biomechanical testing and opening the door for developments in skin research.

14.3.4 Tribological Device to Characterise the Friction Behaviour of Shoes to Prevent Slipping

Human slips and falls are prevalent across the globe in areas such as workplaces, hospitals, homes, and bathrooms [40–43]. These accidents could lead to several fatal and non-fatal injuries such as, muscle tears, ligament tears,

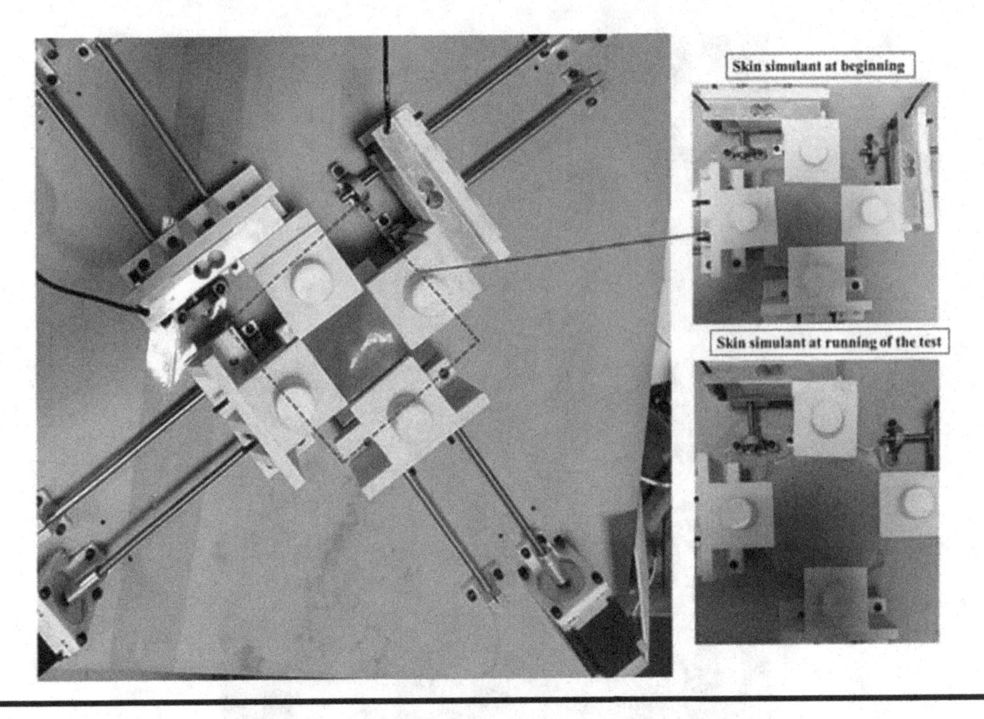

Figure 14.18 Custom developed biaxial tester [38].

fractures, dislocations, or any permanent disabilities [44–49]. Slipping generally occurs due to a reduction in the friction at the shoe-floor contact, leading to instant slippage [50–54]. Hence, tribometers which could judge the material friction or tread-based design friction are important to investigate the quality of footwear.

Gupta et al. [44] recently presented the development of a tribometer which could effectively measure the friction of shoes based on the biomimicry of slipping mechanics. Figure 14.19 represents the developed tribometer to ascertain shoe friction.

In summary, the built low-cost slip testing apparatus met the biomechanical requirements of previous research; that is, it could duplicate the average weight, slipping speed, and heel angle found in human slipping experiments. Their innovative biofidelic traction testing device is intended to have an influence on the traction performance testing of slip-resistant and non-slip-resistant footwear in clinical, industrial, and commercial environments. Employees in hospitals, enterprises, and workplaces may benefit from this to reduce the risk of slips and falls. The device's low cost, light weight, and portability, as well as its provisions for workers' compensation related to slips and falls, will allow for field testing of specialised footwear on realistically contaminated floorings such as greasy floors in the manufacturing,

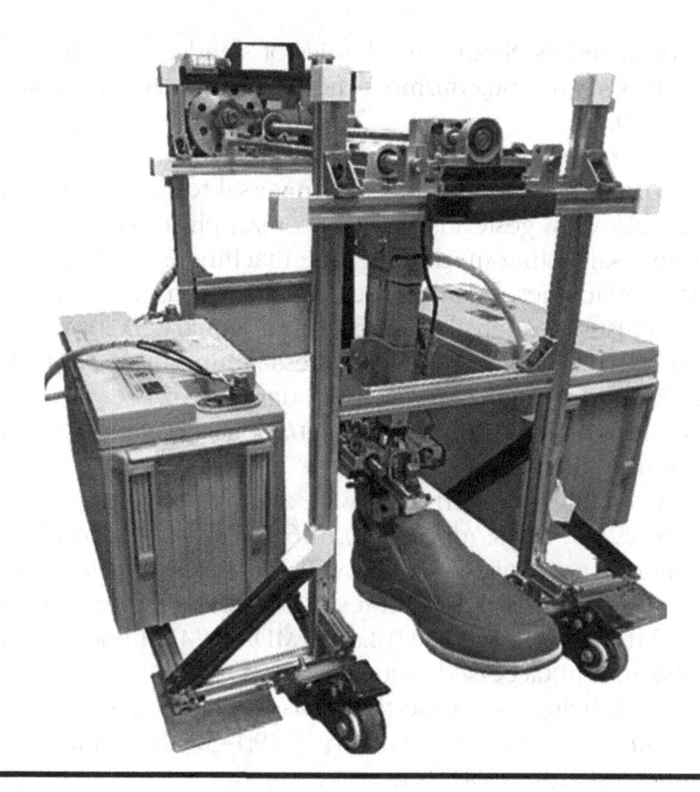

Figure 14.19 Tribometer to ascertain shoe friction [55].

heavy oil, and gas industries, wet flooring after cleaning in commercial buildings, hospitals, and multinational corporations, and oil and food spills on hotel kitchen floors.

References

1. Saba N, Jawaid M, Sultan MTH. An overview of mechanical and physical testing of composite materials. *Mech Phys Test Biocomposites Fibre-Reinf Compos Hybrid Compos* January 2019:1–12. https://doi.org/10.1016/B978-0-08-102292-4.00001-1.
2. Cai C, Zhou K. Metal additive manufacturing. *Digit Manuf Ind "Art to Part" 3D Addit Print* January 2022:247–98. https://doi.org/10.1016/B978-0-323-95062-6.00005-X.
3. Tensile testing - An overview | ScienceDirect topics. https://www.sciencedirect.com/topics/engineering/tensile-testing (accessed January 5, 2024).
4. Mechanical testing – Importance, uses, properties. https://www.azom.com/article.aspx?ArticleID=21909 (accessed January 5, 2024).
5. Pfeifer M. Degradation and reliability of materials. *Mater Enabled des* January 2009:161–87. https://doi.org/10.1016/B978-0-7506-8287-9.00006-9.

6. Mechanics of materials: Strain › mechanics of slender structures | Boston University. https://www.bu.edu/moss/mechanics-of-materials-strain/ (accessed January 5, 2024).
7. Davis JR, ed. *Tensile Testing*. ASM International; 2004.
8. Application, usage and maintenance of universal testing machine tensile machine. https://www.gesterinstruments.com/application-usage-and-maintenance-of-universal-testing-machine-tensile-machine (accessed January 5, 2024).
9. Tensile testing machines | An introduction | Instron. https://www.instron.com/en/resources/test-types/tensile-test (accessed January 5, 2024).
10. Yield strength - Definition, examples , stress-strain graph, FAQs. https://byjus.com/physics/yield-strength/ (accessed January 5, 2024).
11. Torsion test. https://www.testresources.net/applications/test-types/torsion-test (accessed January 5, 2024).
12. Wright RN. Mechanical properties of wire and related testing. *Wire Technol* January 2016:129–57. https://doi.org/10.1016/B978-0-12-802650-2.00011-X.
13. Torsion testing - An overview | ScienceDirect topics. https://www.sciencedirect.com/topics/engineering/torsion-testing (accessed January 5, 2024).
14. Torsion test. https://www.mem50212.com/MDME/MEMmods/MEM23061A/Torsion/Torsion.html (accessed January 5, 2024).
15. Khan Y. Characterizing the properties of tissue constructs for regenerative engineering. *Encycl Biomed Eng* January 2019;1–3:537–45. https://doi.org/10.1016/B978-0-12-801238-3.99897-0.
16. What is torsion testing? | Instron. https://www.instron.com/en/resources/test-types/torsion-test (accessed January 5, 2024).
17. Torsion test - STEP lab. https://step-lab.com/torsion-test/ (accessed January 5, 2024).
18. Mechanical and durability testing of aerospace materials. *Introd Aerosp Mater* January 2012:91–127. https://doi.org/10.1533/9780857095152.91.
19. What is fatigue testing? https://www.wmtr.com/What_Is_Fatigue_Testing.html (accessed January 5, 2024).
20. Creep test | Instron. https://www.instron.com/en/resources/glossary/creep-test (accessed January 5, 2024).
21. Hardness testing: definition, how it works, types and benefits | Xometry. https://www.xometry.com/resources/materials/hardness-testing/ (accessed January 5, 2024).
22. Rockwell hardness tester - Qness 150 CS ECO by QATM. https://www.qatm.com/products/hardness-testing/rockwell-hardness-tester/qness-150-cs/ (accessed January 5, 2024).
23. Buy Vickers hardness testers at best price, Vickers hardness testers manufacturer in Maharashtra. https://www.krystalmachines.com/vickers-hardness-testers-5987313.html (accessed January 5, 2024).
24. Durometer tester AD-100. https://www.checkline.com/product/AD-100 (accessed January 5, 2024).
25. Tribometer - About tribology. https://www.tribonet.org/tribometer/ (accessed January 5, 2024).

26. Randall NX. Experimental methods in tribology. *Tribol Sci Eng From Basics to Adv Concepts* July 2013;9781461419457:141–75. https://doi.org/10.1007/978-1-4614-1945-7_4/COVER.
27. Discover nondestructive testing. https://www.asnt.org/MajorSiteSections/About/Discover_Nondestructive_Testing.aspx (accessed January 5, 2024).
28. What is non-destructive testing (NDT)? Methods and definition - TWI. https://www.twi-global.com/technical-knowledge/faqs/what-is-non-destructive-testing#WhatarethеAdvantagesofusingNDT (accessed January 5, 2024).
29. EWGAE codes for acoustic emission examination: Code IV — Definition of terms in acoustic emission Code V — Recommended practice for specification, coupling and verification of the piezoelectric transducers used in acoustic emission. *NDT Int* August 1985;18(4):185–94. https://doi.org/10.1016/0308-9126(85)90067-7.
30. Magnaflux Limited. Processing line permits fully automatic penetrant testing. *NDT Int* April 1988;21(2):91. https://doi.org/10.1016/0308-9126(88)90419-1.
31. Singh G, Chanda A. Mechanical properties of whole-body soft human tissues: a review. *Biomed Mater* October 2021;16(6):062004. https://doi.org/10.1088/1748-605X/AC2B7A.
32. Ghajar J. Traumatic brain injury. *Lancet* September 2000;356(9233):923–9. https://doi.org/10.1016/S0140-6736(00)02689-1.
33. Cunningham AS *et al.* Physiological thresholds for irreversible tissue damage in contusional regions following traumatic brain injury. *Brain* August 2005;128(8):1931–42. https://doi.org/10.1093/BRAIN/AWH536.
34. Singh G, Chanda A. Development and mechanical characterization of artificial surrogates for brain tissues. *Biomed Eng Adv* June 2023;5:100084. https://doi.org/10.1016/J.BEA.2023.100084.
35. Cheng G, Shan X. Dynamics analysis of a parallel hip joint simulator with four degree of freedoms (3R1T). *Nonlinear Dyn* December 2012;70(4):2475–86. https://doi.org/10.1007/S11071-012-0635-4/FIGURES/6.
36. Hua Z *et al.* Wear test apparatus for friction and wear evaluation hip prostheses. *Front Mech Eng* April 2019;5:440975. https://doi.org/10.3389/FMECH.2019.00012/BIBTEX.
37. Gupta S, Gupta V, Chanda A. Biomechanical modeling of novel high expansion auxetic skin grafts. *Int J Numer Method Biomed Eng* 2022:e3586. https://doi.org/10.1002/CNM.3586.
38. Gupta V, Singla R, Singh G, Chanda A. Development of soft composite based anisotropic synthetic skin for biomechanical testing. *Fibers* June 2023;11(6):55. https://doi.org/10.3390/FIB11060055.
39. Yi D *et al.* Differences in functional brain connectivity alterations associated with cerebral amyloid deposition in amnestic mild cognitive impairment. *Front Aging Neurosci FEB* February 2015;7:110840. https://doi.org/10.3389/FNAGI.2015.00015/BIBTEX.
40. Gupta S, Chatterjee S, Malviya A, Singh G, Chanda A. A novel computational model for traction performance characterization of footwear outsoles with horizontal tread channels. *Computation* February 2023;11(2):23. https://doi.org/10.3390/COMPUTATION11020023.

41. Gupta S, Chanda A. Biomechanical modeling of footwear-fluid-floor interaction during slips. *J Biomech* July 2023;156:111690. https://doi.org/10.1016/J.JBIOMECH.2023.111690.
42. Chatterjee S, Gupta S, Chanda A. Barefoot slip risk in Indian bathrooms: a pilot study. *Tribol Trans* July 2022:1–15. https://doi.org/10.1080/10402004.2022.2103863.
43. Malviya A, Gupta S, Chatterjee S, Chanda A. Development of a novel biomedical device for shoe traction safety characterization. *J Eng Res* August 2023. https://doi.org/10.1016/J.JER.2023.08.018.
44. Gupta S, Malviya A, Chatterjee S, Chanda A. Development of a portable device for surface traction characterization at the shoe-floor interface. *Surfaces* December 2022;5(4):504–20. https://doi.org/10.3390/SURFACES5040036.
45. Gupta S, Sidhu SS, Chatterjee S, Malviya A, Singh G, Chanda A. Effect of floor coatings on slip-resistance of safety shoes. *Coatings* October 2022;12(10. https://doi.org/10.3390/COATINGS12101455.
46. Gupta S, Chatterjee S, Chanda A. Effect of footwear material wear on slips and falls. *Mater Today Proc* April 2022. https://doi.org/10.1016/J.MATPR.2022.04.313.
47. Chanda A, Gupta S, Chatterjee S *Footwear Traction*. Springer Nature Singapore; 2024.
48. Chanda A, Gupta S, Chatterjee S. Footwear wear and wear mechanisms. In: *Footwear Traction. Biomedical Materials for Multi-functional Applications*. Singapore: Springer; 2024, pp. 105–12. https://doi.org/10.1007/978-981-99-7823-6_11
49. Gupta S, Chatterjee S, Malviya A, Kundu A, Chanda A. Effect of shoe outsole wear on friction during dry and wet slips: a multiscale experimental and computational study. *Multiscale Sci Eng* March 2023:1–15. https://doi.org/10.1007/S42493-023-00089-0.
50. Gupta S, Chatterjee S, Malviya A, Chanda A. Frictional assessment of low-cost shoes in worn conditions across workplaces. *J Bio Tribo Corros* March 2023;9(1):1–13. https://link.springer.com/article/10.1007/S40735-023-00741-0.
51. Gupta S, Chatterjee S, Chanda A. Frictional characteristics of progressively worn footwear outsoles on slippery surfaces. *Tribol Ind* September 2023;45(3):416–30. https://doi.org/10.24874/TI.1434.01.23.05.
52. Chanda A, Gupta S, Chatterjee S. Introduction to slips and falls. In: *Footwear Traction. Biomedical Materials for Multi-functional Applications*. Singapore: Springer; 2024, pp. 1–10. https://doi.org/10.1007/978-981-99-7823-6_1
53. Gupta S, Chatterjee S, Malviya A, Chanda A. Traction performance modeling of worn footwear with perpendicular treads. *Tribol Mater Surf Interfaces* October 2023;17(4):352–62. https://doi.org/10.1080/17515831.2023.2246757.
54. Gupta S, Chatterjee S, Malviya A, Chanda A. Traction performance of common formal footwear on slippery surfaces. *Surfaces* November 2022;5(4):489–503. https://doi.org/10.3390/SURFACES5040035.
55. Gupta S, Chatterjee S, Chanda A. Influence of vertically treaded outsoles on interfacial fluid pressure, mass flow rate, and shoe–Floor traction during slips. *Fluids* February 2023;8(3):82. https://doi.org/10.3390/FLUIDS8030082.

Chapter 15

Regulations and Standards: ISO for Healthcare Domain

15.1 Regulatory Requirements for Healthcare Entrepreneurship

Establishing a healthtech firm is a significant task. Offering a refined resolution to an existing issue is one thing, but managing intricate regulations is quite another. Nothing is more annoying than having a perfectly good company concept fall through because of non-compliance with regulations. For this reason, the best approach to prevent uncertainty and financial loss for a healthcare business is to develop a regulatory checklist as soon as possible [1]. Across the globe, different countries have different regulations according to their needs and requirements. Hence, throughout the chapter, only the United States is considered for the explanation of regulations.

The US healthcare sector is highly regulated, competitive, and profitable. A medical company has to stay on top of a lot of rules and regulations in order to succeed in the market. Hospitals and health systems in the United States are required to fulfil 629 standards from four federal agencies, according to the American Hospital Association. Furthermore, several local and state agencies oversee certain local healthcare-related issues. Furthermore, independent, non-governmental groups like the American Medical Association (AMA) have a say in how the US healthcare system is run. For a business that focuses on technology, navigating this legal maze can be quite

DOI: 10.4324/9781003475309-20

difficult. Knowing the main players and the rationale behind the strict rules system, however, can be beneficial. The following are some of the main American healthcare regulators and what they do [1]:

Centers for Medicare and Medicaid (CMS): This organisation is in charge of monitoring adherence to the majority of health laws, including the Health Insurance Portability and Accountability Act (HIPAA). CMS's main objective is to offer Medicare, Medicaid, and State Children's Health Insurance Program (SCHIP) beneficiaries subsidised medical care.

Centers for Disease Control and Prevention (CDC): The main responsibility of the CDC is to keep an eye on the health risks posed by infectious diseases, as the name would imply. But the group also works on birth defects, the environment, disaster relief, and even studies of violence and injury.

US Food and Drug Administration (FDA): Oversees a broad range of laws pertaining to the authorisation of prescription medications, immunisations, dietary supplements, and cosmetics. Remarkably, it also has regulatory jurisdiction over cell phones, other types of medical gadgets, and matters pertaining to disease control.

Department of Health and Human Services: It is a government agency that houses numerous healthcare regulatory bodies, such as the Agency for Healthcare Research and Quality(AHRQ), in addition to the ones mentioned above. The latter carries out studies to raise the standard of US healthcare generally and patient safety in particular.

Environmental Protection Agency: It develops and implements laws meant to safeguard the environment and public health.

Adherence to regulatory requirements in the healthcare industry is crucial. They uphold greater standards of care, safeguard the rights of patients and medical personnel, and guarantee that any newly released goods are legal and safe. Regulating a digital product does not absolve it of its obligations. Within the United States, approval from the FDA is mandatory for all hardware and the majority of software solutions designed for application in the healthcare sector. Adherence to FDA standards is crucial, and it is vital to acknowledge that meeting additional regulatory requirements is necessary. Consequently, the startup must actively pursue approval from this agency [1].

As healthcare services are directed towards the patient, the application will likely gather, or process sensitive personal information or data. Stringent regulations in the United States pertaining to software in healthcare underscore the utmost importance of safeguarding the safety and security of patient data. Regulatory agencies are vigilant in ensuring the protection of patient information at all times.

Unforeseen events may occur, requiring legal readiness for the most unfavourable outcomes. If your software incorporates diagnostic features or offers therapeutic guidance, it introduces the potential for negligence or patent claims. The most effective means of averting future legal disputes involves strict compliance with applicable standards and the establishment of a strong legal framework. It is imperative to integrate thorough legal planning into the project's initial phases.

In the absence of addressing regulatory compliance concerns early in the development process, a bottleneck may arise, potentially obstructing or entirely impeding the progress of the product at a later stage. Consider the financial implications of code modifications as the project nears completion, or envisage the complexities associated with obtaining last-minute certification shortly before the scheduled launch. It is pertinent to acknowledge that the regulatory approval process tends to be time-consuming with most regulatory agencies. Regulations demanding attention encompass [1]:

FFDCA: The Federal Food, Drug, and Cosmetic Act encompasses a set of laws and regulations that govern the development and usage of pharmaceuticals, medical devices, and various other entities. A comprehensive understanding of the specific laws pertaining to the startup's focus is crucial for the legal department, given the multitude of regulations.

In essence, interfacing with the FDA is inevitable in achieving regulatory compliance for a healthcare venture. Such engagement is indispensable when striving to establish regulatory adherence for the healthcare startup [1].

HIPAA and HITECH Act: The Health Insurance Portability and Accountability Act of 1996 serves as a fundamental safeguard for customer personal data throughout its entire journey. Its provisions delineate rules governing the collection, storage, exchange, and disclosure of such data. Compliance with HIPAA requirements is essential for healthcare startups, especially those engaged in projects related to processing protected health information (PHI). This encompasses the development of electronic medical records (EMRs) or electronic health records (EHRs) for medical facilities and software for use by their business associates, including insurers, law firms, and others. Strict adherence is crucial to pre-empt legal complications.

The Anti-Kickback Statute and Stark Law: Establishing fundamental rules is imperative when formulating the ethics code for a startup. The primary objective is to incorporate provisions that prevent bribery and fraud in any manifestation. Adhering to the Stark Law, also known as the Ethics in Patient Referrals Act or Physician Self-Referral Law, is crucial to preclude physicians from obtaining personal benefits when referring patients to healthcare providers. Simultaneously, the Anti-Kickback Statute treats such actions as criminal offenses, imposing penalties such as fines reaching $25,000 or up to five years of incarceration.

In cases involving telemedicine and/or virtual care, it is essential to distinctly delineate lawful boundaries for incentivising both patients and medical professionals. Additionally, startups must be cognisant of specific and local regulations. For instance, operating in California necessitates compliance with the California Consumer Privacy Act (CCPA), which surpasses most HIPAA requirements. Modelled closely after the European Union's General Data Protection Regulation (GDPR), the CCPA enforces stringent standards for the safety of personal data.

15.2 Cybersecurity Regulation Frameworks for Healthcare Ventures

In 2021, the healthcare sector incurred the highest average total cost of data breaches among all industries, amounting to $9.23 million [2]. Additionally, the exposure or theft of 44,993,618 health records marked 2021 as the second-highest year for breached records (Figure 15.1) [2]. The advent of digital transformation elevates the vulnerability of all industries to cyber threats, and the healthcare sector is no different. As US healthcare entities

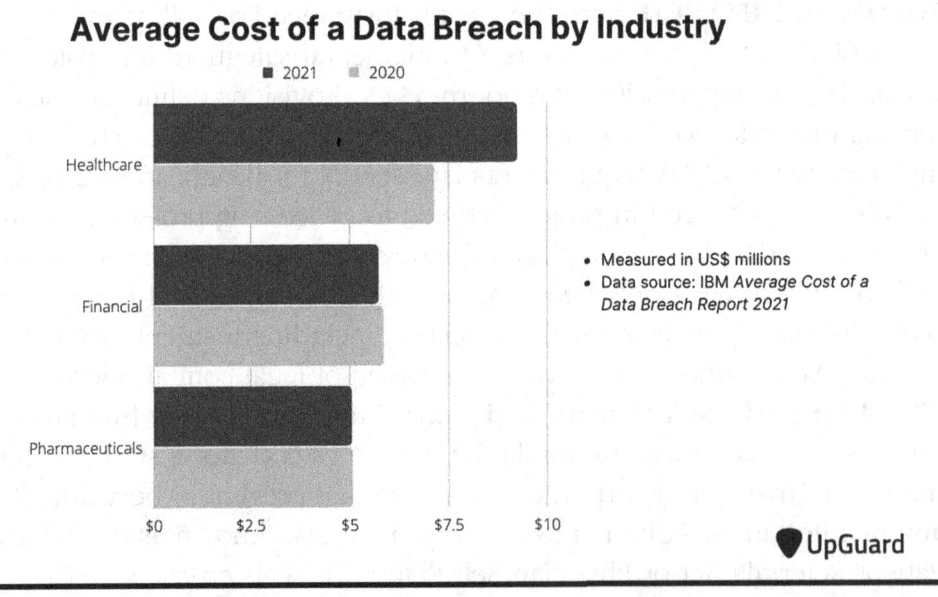

Figure 15.1 Average cost of data breach across different industries [2].

increasingly rely on health information technology for functions such as data sharing, process automation, and system interoperability, the potential points of attack multiply rapidly. This proliferation of attack vectors significantly heightens the overall cybersecurity risk. Within the healthcare industry, particular susceptibility is observed in the realms of ransomware attacks, data theft, and endpoint compromise.

Entrances for ransomware attacks can be as simple as infiltrating a system through methods like phishing emails. The healthcare sector faces heightened risks, as data loss in a patient care facility not only results in inconvenience but also poses a direct threat to patient safety. Cybercriminals capitalise on this urgency, using it as leverage to demand larger sums of money

Compliance with healthcare regulations not only helps organisations avoid hefty fines but also contributes to a more consistent and measurable improvement in the maturity of their security posture. Obtaining certification through recognised security frameworks enhances organisational credibility and facilitates the assessment of regulatory compliance. These frameworks streamline the labour-intensive task of creating a cybersecurity roadmap from scratch.

Merely adhering to the framework requirements allows firms to assess their security posture and identify areas of compliance and non-compliance when there are financial or time constraints. A clear trajectory toward security posture maturity and the thorough handling of capacity gaps are ensured by a well-defined strategy for establishing cyber resilience. The following are eight prominent cybersecurity regulations and frameworks that US healthcare organisations should prioritise when formulating their information security policies [2].

National Institute of Standards and Technology (NIST) Framework: A comprehensive set of industry rules aimed at reducing cyber hazards for enterprises is the National Institute of Standards and Technology (NIST) Cybersecurity Framework. NIST 800-53 is one of the many cybersecurity publications that NIST has published. NIST 800-53 was once developed to create security and privacy guidelines that applied only to federal and government organisations. However, the most recent modification (Revision 5) expands its application to non-government organisations, such as the healthcare industry. With this update, security and privacy controls are combined to create a single, cohesive set of controls that can be used by organisations and systems alike.

Moreover, NIST simplifies compliance complexities by seamlessly integrating with other frameworks and regulations, such as HIPAA and ISO 27001. This integration minimises challenges associated with meeting diverse compliance requirements, allowing for a more streamlined and efficient cybersecurity approach.

Health Insurance Portability and Accountability Act (HIPAA): HIPAA, a set of US federal laws enacted in 1996, aims to oversee the disclosure and protection of health information within the nation. Comprising three key rules—Privacy Rule, Security Rule, and Breach Notification Rule—HIPAA establishes comprehensive guidelines.

One essential element, the Privacy Rule, outlines the conditions under which covered entities may use or divulge a person's health information. Protected health information (PHI), including electronic health protected information (ePHI), cannot be disclosed by covered entities unless the Privacy Rule specifically permits it or unless the person providing the information or their agent authorises it in writing. Exceptions for mandatory PHI disclosure include situations where an individual or their representative requests access to it or when the Department of Health and Human Services (HHS) is involved in a compliance investigation, review, or enforcement action.

Covered entities which do to comply with the law may be subject to civil penalties enforced by the Department of Health and Human Services' Office for Civil Rights (OCR). There is an annual cap of $1,500,000 and penalties ranging from $100 to over $50,000 for each infraction. In addition to civil penalties, certain Privacy Rule violations could also result in criminal prosecution. This underscores the gravity of non-compliance under HIPAA, indicating the potential for both civil and criminal consequences for entities found in violation.

Center for Internet Security (CIS) Security Controls: CIS has developed the critical security controls to fortify both private and public organisations against cybersecurity threats. These controls, totalling 18 (previously 20), are essential for protecting organisations from cyberattacks. They encompass a wide range of actions, such as data protection, secure configuration of software and enterprise assets, inventory and control of software assets, malware defences, data recovery, network infrastructure management, incident response management, security awareness and skills training, service provider management, application software security, penetration testing, and continuous vulnerability management.

15.3 ISO Standards for Healthcare Ventures

The healthcare industry encompasses various entities dedicated to ensuring people's health and safety, such as drug manufacturers, hospitals, pharmacies, and medical device industries. Maintaining high standards in product and service quality, as well as operational efficiency, is crucial in this sector. To demonstrate credibility in the market, the International Organization for Standardization (ISO) publishes a specific set of certifications tailored for the healthcare industry. These certifications serve as benchmarks for excellence and efficiency within the healthcare sector. The following major standards could be explored before starting a healthcare startup [3].

ISO 9001: By implementing a quality management system (QMS) based on the ISO 9001 standard, organisations can facilitate periodic management reviews and internal audits. These processes are instrumental in identifying any shortcomings in the system, allowing the organisation to apply corrective measures and ensure the effectiveness of its QMS. The emphasis on risk management and continual improvement further enables organisations to meet evolving market demands, enhancing their adaptability and overall performance.

ISO 14001: The implementation of the Environmental Management System (EMS) in organisations, guided by the ISO 14001 standard, contributes to reducing the environmental impacts within the healthcare sector. Through EMS, healthcare organisations can effectively manage and minimise wastage, ensuring compliance with environmental laws. This proactive approach not only helps in reducing the environmental footprint but also establishes a reputation in the market for environmental responsibility, enhancing the organisation's standing and credibility.

ISO 45001: The significance of ISO 45001 certification becomes evident in ensuring the occupational safety of healthcare workers. This certification plays a crucial role in preventing healthcare hazards by timely identification of risks and implementing preventive measures. By adhering to ISO 45001 standards, healthcare organisations can create a safer work environment, reducing the likelihood of occupational risks and contributing to the well-being of their workforce.

ISO 27001: Healthcare organisations bear the responsibility of safeguarding personal information about their patients or customers. The protection of customer privacy is a crucial aspect entrusted to these organisations. Consequently, the adoption of the ISO 27001 standard for Information Security Management System (ISMS) becomes instrumental. This implementation ensures that organisations in the healthcare sector take comprehensive measures to protect data, mitigating the risk of breaches or losses and upholding the confidentiality and security of sensitive information.

ISO 13485: This standard is instrumental in instituting QMSs for medical devices, emphasising the importance of safety. Whether the device is external or implantable, its safety is critical, directly influencing patient well-being. By adhering to this standard, organisations ensure that medical devices meet high quality standards, guaranteeing their premium quality and complete safety for use in healthcare settings.

ISO 13485 holds immense significance in the medical device sector, providing manufacturers with a framework to consistently meet regulatory requirements and fulfil customer expectations [4]. Compliance with ISO 13485 is pivotal for companies seeking regulatory approval across diverse markets, thereby boosting competitiveness and expanding market access.

The benefits of ISO 13485 encompass enhanced product quality and safety, fostering the establishment and maintenance of robust quality systems. This compliance builds customer confidence by demonstrating the implementation of a reliable QMS. Recognised by regulatory bodies globally, ISO 13485 serves as credible evidence of adherence to medical device production requirements.

Moreover, ISO 13485 facilitates entry into international markets, showcasing commitment to meeting rigorous global quality standards. The standard also promotes continual improvement by mandating the ongoing monitoring and optimisation of processes, contributing to increased efficiency and reduced costs over time. In essence, ISO 13485 is indispensable for medical device manufacturers, ensuring the quality and safety of products and facilitating seamless entry into the global market [4].

ISO 50001: Healthcare companies have the opportunity to decrease operational expenses and enhance overall efficiency through the successful implementation of ISO 50001 for energy management systems. This standard provides a framework for organisations to systematically manage and improve their energy performance, enabling healthcare entities to optimise energy usage, reduce costs, and contribute to sustainable practices.

In addition to these, ISO has specialised committees which constantly work in the healthcare domain to create new and updated standards. These fields include dentistry, assistive products, optics, sterilisation of healthcare products, transfusion equipment, healthcare informatics, infusions and injections, medical devices, surgical implants, surgery, and healthcare organisation management. The following few major committees should be studied, in addition to the standards, to fully understand the individual requirements based on the healthcare product or service [3, 5]

ISO/TC 210—Medical device quality control and related general aspects: The health committee in question is responsible for developing guidelines regarding the use of risk management in the creation and manufacturing of medical devices. It also covers the standards for small-bore connections that are used to link different medical equipment, such as syringes, tubing, and other accessories that are used to transfer fluids and gases for patient care. This committee addresses the compatibility and appropriate operation of interconnected components in healthcare settings and is crucial in developing standards to guarantee the safety and efficacy of medical devices.

ISO/TC 212—Systems for in vitro diagnostic testing and clinical laboratory testing: The aforementioned health committee provides a practical approach to reduce errors in medical laboratories by implementing risk management and continuous improvement procedures. Current norms falling under its jurisdiction include quality control, analytical performance, and laboratory safety. The committee is also geared towards future areas of emphasis, particularly in the management of biorisks. These biorisks involve risks stemming from the handling of biological agents and toxins, reflecting the committee's commitment to addressing evolving challenges in the field of medical laboratory practices.

ISO/TC 194—Clinical and biological assessment of medical devices: The primary goal of the ISO technical committee is to standardise biological test procedures that are used to evaluate dental and medical materials and equipment prior to their release onto the market. This group is essential in setting common standards and requirements that guarantee a comprehensive assessment of these materials and devices, enhancing their overall quality, safety, and efficacy prior to going on sale.

ISO/TC 198—Sterilisation of medical supplies: In order to ensure the successful sterilisation of medical supplies, the relevant health committee lays out the requirements for sterilising procedures, sterilising apparatus, washer disinfectors, and related goods. Its function includes putting in place strict guidelines to guarantee the complete and dependable sterilisation of different medical supplies, enhancing the general security and effectiveness of medical procedures.

ISO/TC 172—Photonics and optics: The health committee works to ensure that safety risks in the relevant domain are addressed in a uniform and widely understood manner by contributing to the establishment of standardised vocabulary and test procedures. This standardisation encourages a methodical and efficient management of safety-related concerns while supporting communication that is clear and precise.

ISO/TC 106—Dental related: Healthcare technology, especially in the area of oral healthcare, is a significant industry within ISO. Standardisation is the main focus of this industry, which includes testing procedures, terminology, and specifications that apply to a variety of supplies, tools, appliances, and equipment used in all areas of dentistry. The objective is to provide standardised rules that will improve the safety, quality, and uniformity of oral healthcare procedures and supplies.

ISO/TC 150—Implants for surgery: The health committee in question provides guidelines and test procedures designed especially to guarantee the efficacy and safety of surgical implants meant to be inserted into the body for medical or diagnostic purposes. This group is essential to the development of standards that improve surgical implant performance and reliability, with the goal of improving patient safety and the overall effectiveness of these medical devices.

ISO/TC 304—Management of healthcare organisations: This health committee is committed to developing standards for the categorisation, nomenclature, measurements, management techniques, and best practices that apply to the non-clinical operations of healthcare organisations. It also incorporates ideas from scientific research. The group is now working hard to produce three standards that cover hand hygiene procedures (ISO 23447), patient-centred staffing (ISO 22956), and the lexicon of healthcare management (ISO 22886). These standards aim to establish guidelines that enhance the efficiency, clarity, and patient-centric focus of non-clinical healthcare operations.

ISO/TC 215—Health information technology: The health committee works to ensure that disparate systems can use health-related data, information, and knowledge in a way that is compatible and interoperable. Its main objective is to make it easier for these systems to integrate seamlessly, supporting different aspects of the health system. By striving for compatibility and interoperability, the committee aims to enhance the efficiency, coordination, and effectiveness of health-related processes and services.

References

1. Regulatory requirements for healthcare startups: what to know to ensure compliance n.d. https://demigos.com/blog-post/regulatory-requirements-for-healthcare-startups/ (accessed January 6, 2024).
2. Top 8 healthcare cybersecurity regulations and frameworks | UpGuard n.d. https://www.upguard.com/blog/cybersecurity-regulations-and-frameworks-healthcare (accessed January 6, 2024).
3. ISO - international organization for standardization n.d. https://www.iso.org/home.html (accessed January 6, 2024).
4. ISO 13485 and it's significance in the medical devices domain n.d. https://blog.decos.com/en/webinars/iso13485 (accessed January 6, 2024).
5. Internal Standards Organization. ISO and health n.d. https://www.iso.org/files/live/sites/isoorg/files/store/en/PUB100343.pdf (accessed January 6, 2024).

SALES, MARKETING, AND COMMERCIALISATION

Chapter 16

Introduction to Branding and Brand Awareness

16.1 Introduction

The process of branding encompasses the development of a distinct name, logo, and image for a specific product, service, or company. The primary objective is to draw in customers, typically achieved through advertising that maintains a uniform theme [1]. The goal of branding is to create a notable and unique presence in the market, fostering the attraction and retention of loyal customers. A brand, whether in the form of a name, term, symbol, or other feature, serves to differentiate an organisation or product from its competitors in the eyes of the customer. The utilisation of brands is prevalent in business, marketing, and advertising [2].

Strategic planning of branding should align with the specific target audience, as it is impractical for any business entity to cater to the entire population. Business proprietors must discern the demographic characteristics of individuals purchasing their products and services. Thorough research is imperative, encompassing factors such as age, gender, income, and the lifestyle of their clientele. Established companies enjoy the advantage of significant brand awareness. To enhance brand visibility, advertising on platforms such as television, radio, newspapers, and social media proves instrumental [2]. Logos play a crucial role in bolstering brand recognition, as individuals frequently identify brands through these distinctive symbols or graphical representations.

DOI: 10.4324/9781003475309-22

The pinnacle of success for any company lies in achieving brand loyalty. A customer extensively purchasing products from a specific company is termed a brand loyalist. Numerous consumers exhibit a preference for particular brands of clothing, deodorants, toothpaste, and more, drawn to the perceived benefits offered by these brands. Establishing brand loyalty involves maintaining regular communication with customers and soliciting their feedback to foster a strong connection. Consistency is imperative for a brand's success, requiring steadfast adherence to a unified image. Small businesses often articulate promises in commercials and advertisements regarding their brands, and consumers anticipate ongoing fulfilment of these commitments. In addition to upholding promises, the efficacy of their products is paramount. Now, some insights into brand awareness should also be considered.

Brand awareness, also termed brand recognition, indicates the degree to which a customer is familiar with a product or service by its name. It signifies the initial phase of a consumer's interest in a product or service and serves as the initiation of their interaction with a brand in the journey towards making a purchase [3]. One might contemplate why recognition, recall, and the establishment of an association with a company occur, even if the consumer does not utilise the company's products [4]. This occurrence is a result of a robust brand awareness, denoting that consumers are acquainted with or have awareness of the company's brands.

A business's brand encompasses more than its logo or tagline; it comprises the products offered, storytelling approach, aesthetic, the delivered customer experience, the company's values, and various other elements. Consider a close associate as an analogy. The initial impression they made during your first encounter left an impact [5]. Subsequent interactions shaped perceptions over time, resulting in a comprehensive understanding of their identity and values. In this perspective, your best friend possesses a brand, amalgamated from all shared experiences. A customer's interpretation of a brand is forged through a spectrum of inputs accumulated over time. Companies construct brands by consistently conveying a unified message and delivering a consistent experience across various touchpoints. This uniformity, characterised by the repetition of messaging and experience, is pivotal in ensuring the memorability of a brand, a cornerstone in cultivating brand awareness.

Brand awareness plays a pivotal role in ensuring that a brand is prominently considered by potential customers when making purchase decisions [6]. While possessing a robust brand is important, expanding a business requires ensuring that consumers are aware of it [7]. The significance of brand awareness lies in its ability to instil trust and afford brands the opportunity to convey their narrative and establish credibility with consumers [8].

According to a 2022 global survey conducted by Statista, half of consumers expressed a willingness to pay a premium for a brand with an appealing image [8]. The Statista report further notes that in 2022, the combined value of the world's top 100 most valuable brands surged by over 22%, reaching a record $8.7 trillion [8]. This represents a substantial increase from approximately $5 trillion just two years prior. Brand awareness also holds importance in shaping a distinctive identity through which a company can communicate its values and mission. This form of connection is particularly significant for consumers. According to a 2022 Amazon Ads and Environics Research report, 79% of global consumers indicate a higher likelihood of purchasing from brands whose values align with their own. Now, understanding branding with respect to healthcare is important and the main goal of this chapter. The explanation is as follows.

16.2 Branding in the Healthcare Domain

The process of creating a unique brand for a healthcare organisation with the aim of influencing the opinions of both potential and existing patients as well as the general public is known as healthcare branding. This extends beyond the design of taglines, logos, or distinctive brand messaging. Instead, healthcare branding entails creating a deep bond between the people the business serves and its vision, mission, and values. This relationship reaches out to local and international communities in addition to patients and their families [9].

Fostering patient acquisition, contentment, and loyalty in the setting of healthcare institutions requires the development of a brand identity through strategic marketing and unique patient encounters. In contrast to other sectors, the healthcare provider–patient relationship's inherent trust can be quickly compromised or destroyed by unpleasant encounters, which can damage the brand's reputation.

Patients' choice of healthcare providers is largely influenced by the crucial role that compassion and care play, with 60% of respondents rating it as "very significant" when making their choice. [9]. In a 2021 Deloitte survey, 55% of respondents indicated that their trust in healthcare providers had been compromised by negative experiences [9]. Furthermore, four out of five respondents asserted that, following such an experience, nothing could persuade them to revisit the same provider. The repercussions extend beyond the provider, as 36% of patients opted to skip or avoid healthcare options due to dissatisfaction with their treatment by a provider [9].

The demographic targeted by healthcare brands seeks many of the same attributes expected from companies in various aspects of their lives. While considerations such as convenience, easy access, and affordability are important, they alone are insufficient. Empathy, the provision of quality care, and the establishment of a trustworthy relationship are paramount [10]. In the realm of healthcare, transactions are not merely transactional; rather, the emphasis is on developing trust and demonstrating genuine care—how the brand is perceived holds significance. Therefore, the brand strategy becomes an indispensable tool for both attracting and retaining patients. This strategic approach must align seamlessly with delivering on the brand promise in tangible patient experiences and how the brand is perceived within the community.

The ability to promptly address challenges in brand perception, a possibility even for the most well-established healthcare brands, is essential to mitigate any impact on success resulting from a tainted reputation. Additionally, a brand strategy assumes a crucial role in catalysing new patient acquisition and broadening market share by motivating patients to transition to a specific location, provider, or service.

16.3 The Advantages of Building a Powerful Healthcare Brand

Increasing share in market: Opportunities to appeal to underrepresented or emerging sectors exist as the US population and local demographics change. For example, it is predicted that by 2040, the number of Americans 65 and older will have increased by 44% [9]. The consumer base grows as a result of an ageing population using healthcare services more frequently. But being able to reach a wider audience depends on having a well-known brand and an excellent reputation [11].

Improving the acquisition of new patients using internet reviews and word-of-mouth advertising: roughly 75% of potential patients base their decisions on unfavourable comments, and 80% of them depend on internet reviews when choosing a new healthcare practitioner [9]. Building a true

Good afternoon

Welcome to Humana, where healthcare centers around you

See how Humana plans put **you** first.

Explore our plans →

Figure 16.1 Example—Humana.com [9].

consumer connection with your healthcare brand makes it more likely that patients would promote it on social media, in reviews, and in other ways that work well for brand marketing.

Promoting fidelity: Patients want a medical facility that is seen as compassionate and understanding. The brand may create significant moments that resonate with its target audience by coordinating healthcare marketing and patient experience with the delivery of this commitment. Adapting to evolving patient demands and market prospects: there is a chance to pinpoint patient needs and spot developing chances by closely observing and examining consumer comments about the brand, both online and in the waiting area.

Improving the experience quality of a patient: It is possible to dramatically increase brand perception, customer lifetime value, and loyalty by deeply understanding the intent and emotions of patients at every point of contact and designing brand experiences that align with this thorough understanding.

The following are some healthcare branding examples from different websites.

- Humana: The initial impressions of this brand benefit from the name “Humana,” signifying a prioritisation of a human-centric approach. The brand’s website greets visitors with “Good morning” or “Good afternoon,” depending on the time, and asserts, “Welcome to Humana, where healthcare centers around you” (Figure 16.1). This promptly

Figure 16.2 Example—Planned Parenthood [9].

Figure 16.3 Example—Medtronic [9].

shapes consumers' perceptions of the brand in a positive manner. In 2021, Humana ranked as the world's third most valuable healthcare brand, underscoring the success of this approach.

- Planned Parenthood: The slogan "Care. No matter what." denotes the brand's commitment to assisting all individuals in need, emphasising the organisation's not-for-profit orientation (Figure 16.2). The organisation's straightforward style and action-oriented language, such as "Get Care" and "Get Involved," align with its social justice and advocacy-focused social media channels.
- Medtronic: Medtronic's tagline (Figure 16.3), "Engineering the extraordinary," underscores the technical aspect of the healthcare brand's endeavours, reinforcing the "tronic" element in the brand name. While its primary focus is on developing healthcare technology solutions not directly targeted at consumers, Medtronic reaffirms its commitment to a people-first approach through statements like "Inspiring hope and new possibility in people all over the world." As of 2021, Medtronic was the fifth most valuable healthcare brand globally.

16.4 Healthcare Marketing Techniques

Engaging both hemispheres of the brain—the left hemisphere, which processes social emotions like happiness, and the right hemisphere, which processes primal emotions like fear—is essential to effective branding. Marketing is quite important in the healthcare sector, particularly when it comes to producing movies that are optimised to meet both cognitive needs [12]. Visual storytelling is the key, marrying scientific information with emotional elements to establish a profound connection between the audience and the brand. Whether it's a healthcare brand or a healthcare branding strategy, the ability to forge this emotional connection is fundamental to achieving success.

16.4.1 Convey Warmth and Compassion

Scientific information, being inherently rational and devoid of emotion, may not independently influence market dynamics. In the healthcare sector, successful brands adeptly convey complex scientific concepts to patients using visually engaging images, simple language, and a compassionate approach. For healthcare brands, the strategy involves identifying narratives that exude warmth, humanisation, and compassion, all while showcasing elements of innovation. Whether it's a healthcare professional sharing their defining moment in choosing the profession or a patient recounting a survival story, authentic narratives from real individuals foster a sense of community [12].

Healthcare stories, particularly patient testimonials, are often rich in emotions, encompassing feelings like fear, anxiety, happiness, and gratitude. The infusion of these emotional elements plays a crucial role in establishing trust in healthcare marketing. The meticulous curation process for such content is equally vital, encompassing the development of client relationships, comprehensive market research to comprehend the perspectives of both current and potential patients, a thoughtful selection of diverse stories for sharing, and ensuring individuals feel comfortable as they share their experiences.

16.4.2 Social Connection

Facebook, Instagram, Twitter—what ties them together? The answer lies in the substantial user base they collectively host. Currently, more than 3.5 billion people utilise social media worldwide [12]. Modern healthcare marketing techniques leverage this extensive reach to create responsive campaigns that motivate audiences to take action, thereby enhancing their brand identity and overall strategy [13].

The traditional model of push campaigns has transformed into dynamic, two-way conversations, allowing marketers to provide prospective patients and customers with precisely what they seek, precisely when they seek it [14]. Mastering the art of predicting and controlling recall and recognition of a healthcare brand represents a valuable marketing asset.

The strategic development and distribution of social-optimised video ads emerge as a highly effective method to not only broaden outreach but to pinpoint the right audience. In the fiercely competitive healthcare industry, brands must seize opportunities to foster engagement, share content, and encourage interaction [15]. Crafting social video ads that resonate seamlessly across diverse platforms enables brands to distinguish themselves, paving the way for successful medical educational campaigns. In terms of content that maximises brand awareness, visual images and videos stand out as the top two most accessed types of content in social ads. This underscores the significance of creating compelling visual narratives to captivate and inform the audience effectively.

16.4.3 Developing an Organic Presence and Sharing Stories

Optimising video content for the web has proven to increase the likelihood of a website appearing on the first page of a search by 53 times [12]. This fundamental boost in visibility enhances an organisation's online presence significantly. While the volume of organic searches remains consistent, there is a shift in the nature of the content people are seeking. Creating campaigns with optimised spending that effectively target the right audience in the right manner is essential. To expand the organic presence of a healthcare brand, leveraging educational video content and patient testimonial videos proves to be particularly effective. These approaches not only contribute to brand growth but also align with the evolving needs of online audiences.

Healthcare deals with intricate subjects, often beyond the comprehension of individuals without a background in biochemistry, ourselves included. Consequently, educational video content in healthcare becomes imperative, offering a means to distil complex topics into actionable insights. Video, recognised for its ability to captivate and allow repeated viewing, emerges as the preferred method for education in this context. Combining patient education with engagement not only imparts valuable information but also positions the brand as a leader within the healthcare landscape. The crux lies in crafting educational video content that not only informs but also engages, simplifying intricate information without leaving viewers feeling uninformed. Achieving this balance involves presenting content with sensitivity rather than sentimentality, opting for explanation over self-centeredness. This approach ensures the resonance of the content, effectively reaching and connecting with the audience.

Patient testimonial videos offer a potent means to heighten brand awareness without veering into inappropriate or misleading messaging [16]. These videos become a cornerstone for building credibility and trust by vividly portraying the experiences of patients, both in moments of triumph and adversity, thus bringing the brand to life.

The narratives shared in patient testimonial videos delve into profoundly personal accounts, encompassing stories of resilience that uplift and tales of sorrow that elicit profound empathy. Patients willingly entrust the brand with their intimate experiences, creating a significant responsibility to authentically honour these narratives. Beyond establishing emotional connections for current and potential customers, these videos play a pivotal role in humanising the brand. In the healthcare sector, where the core of the experience revolves around feeling human and respected, patient testimonial videos become a catalyst for creating enduring impressions [12].

Presenting brand messaging in the form of a story increases memorability by up to 22 times compared to relying on facts alone [12]. Beyond information retention, the paramount importance lies in the capacity to forge more profound connections with individuals. Storytelling serves as a transformative tool, rendering healthcare brands more human and relatable, thereby transcending conventional advertising approaches. Visual storytelling, particularly through a cinematic lens, introduces an authentic and visually appealing dimension, elevating healthcare brands beyond traditional advertising norms. Video advertising, in this context, holds the power to reconnect healthcare with the heart of medicine, effectively conveying a genuine sense of care and compassion.

16.4.4 Positioning the Brand

In the dynamically evolving healthcare sector, it is imperative to establish a brand characterised by consistency, reliability, and distinctiveness to stand out amidst competition. Achieving this necessitates comprehensive competitive brand research to identify opportunities for differentiation, involving a profound understanding of the competitive environment within which the brand operates and discerning unique aspects that set it apart [17].

While other providers may offer similar procedures, the pivotal question patients seek answers to revolves around what sets certain providers apart. To attain differentiation and cultivate a robust brand identity, it is crucial to define the target audience, competitive advantage, and brand promise. Addressing these questions purposefully and authentically aids in developing healthcare brand awareness that resonates with patients and distinguishes the brand from competitors [12]. Subsequently, the unique value proposition emerges as the factor that sets the brand apart, catering to the preferences of the target audience. To define it, assess the specific benefits offered to patients, the unique aspects of services or care, and why patients would choose this brand over another provider.

Finally, the brand promise highlights the unique characteristics of the healthcare organisation and communicates what patients may expect from the brand. Think about the values the brand represents and how these fit the requirements and tastes of the intended audience before creating it.

16.4.5 Creating a Corporate Identity

This is a crucial component in building brand recognition in the healthcare industry. Even if healthcare services aren't usually that unique, a strong brand identity is essential for setting a company apart in a crowded and cutthroat market.

To develop a robust brand identity, various strategies can be considered. Crafting a memorable logo stands out as an indispensable aspect. It should possess simplicity, memorability, and reflect the brand's values and personality. Collaboration with a professional designer can facilitate the creation of a logo that not only stands out but is also easily recognisable by patients [17].

Ensuring consistency in the colour scheme and typography across all brand materials, spanning the website to office signage and uniforms, is equally important. Opting for colours and fonts that align with the brand's values and personality resonates effectively with the target audience. Moreover, maintaining consistency in brand messaging is vital, ensuring alignment with the brand's values and personality [18]. Tailoring it to different patient audiences and utilising it across all touchpoints reinforces the brand identity effectively.

16.4.6 Creating a Narrative

An essential element in healthcare brand awareness resides in the practice of storytelling. By imbuing your brand narrative with humanity and depth, an emotional connection with patients is forged, fostering a profound sense of trust and loyalty. One impactful strategy for constructing a narrative around the brand involves sharing stories of patients whose lives have been positively transformed with the assistance of the organisation and its employees. Illustrating the constructive influence the brand exerts on patients nurtures credibility and instils confidence.

These patient narratives can be disseminated through diverse channels, such as social media, blog posts, and patient testimonials, always mindful of adherence to HIPAA regulations. Simultaneously, the narrative should encompass the pivotal role played by employees in shaping the brand story. Spotlighting their expertise, dedication, and commitment to patient care serves to reinforce brand values and the overarching mission. Featuring profiles of providers and employees, acknowledging their achievements and proficiency, and sharing stories that underscore their unwavering commitment to patient well-being all contribute to enhancing the brand narrative [17].

16.4.7 Staying Connected with Patients

With an excess of 191.2 million social media users in the United States, establishing a presence where patients actively engage becomes imperative. Sustaining connectivity with patients on social media involves diverse strategies. Regular posting is indispensable to remain at the forefront of patients' minds, incorporating a diverse content mix encompassing educational content, patient narratives, provider profiles, and posts unrelated to services.

Building relationships and fostering a sense of community around the brand are crucial aspects of social media interaction. Engagement with the audience, through posing questions, responding to comments, and sharing or retweeting content of interest or value, is essential. Authenticity holds paramount importance in establishing a strong connection. Avoiding excessive promotion and focusing on relationship-building and providing value, treating social accounts as avenues for connection rather than purely for business gains, is crucial. Adherence to HIPAA guidelines is essential in these interactions [17].

Facilitating sharing is integral to healthcare brand awareness. Forge meaningful connections by delivering value, building trust, and encouraging engagement, prioritising connection over financial motives. When patients share positive experiences, it enhances brand credibility. Creating informative, engaging, and relevant content incentivises patients to share and amplify the reach of the healthcare organisation. Utilise eye-catching visuals, hashtags, and calls to action to enhance shareability. Integrating social media sharing buttons on websites and blog posts simplifies patient sharing. Paid advertising on social media platforms targets specific demographics, extending content reach and facilitating patient engagement, ultimately contributing to enhanced brand awareness in healthcare.

Consumers exhibit a fourfold higher likelihood of increased purchases from a health insurer and a 3.1-times higher likelihood of choosing the same hospital or medical clinic for future care following a five-star experience

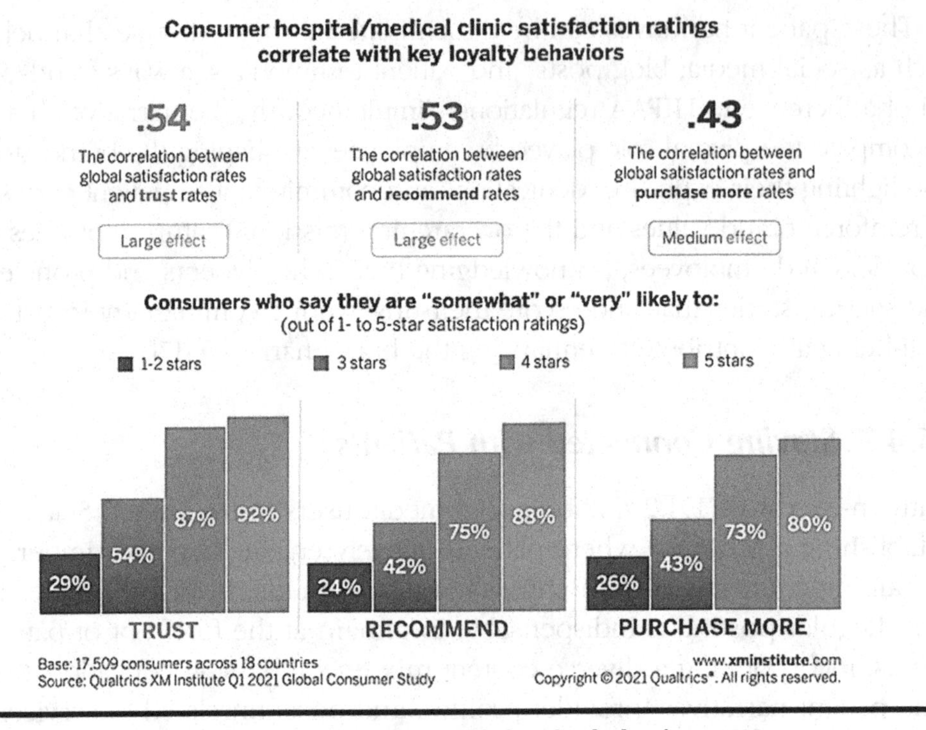

Figure 16.4 Consumer satisfaction ratings with loyalty behaviours [9].

compared to a one to two-star experience [9]. In light of this feedback, refine your brand's experiences to incentivise patients to return for additional services. Figure 16.4 represents the consumer satisfaction ratings with loyalty behaviours. These show high correlation between the satisfaction and trust values. It also makes an impact on the purchasing.

References

1. Keesling G. Brand name changes help health care providers win market recognition. *Health Mark Q* 1993;10:41–53. https://doi.org/10.1300/J026V10N03_04.
2. Branding: introduction, features of branding, concepts, solved examples n.d. https://www.toppr.com/guides/business-studies/marketing/branding/ (accessed January 7, 2024).
3. Fong CH, Goh YN. Why brand equity is so important for private healthcare? View from an emerging market. *Int J Healthc Manag* 2021;14:1198–205. https://doi.org/10.1080/20479700.2020.1755811.
4. Master the future of healthcare branding with our comprehensive guide: discover effective strategies and tools for success — the branded agency n.d. https://www.brandedagency.com/blog/branding-marketing-healthcare (accessed January 7, 2024).

5. Brand awareness - definition, importance, strategy, & examples n.d. https://www.feedough.com/brand-awareness-a-comprehensive-guide/ (accessed January 7, 2024).
6. Pol H, van der Herberg E, Barten DJ, Tielen J, van der Veen G. Brand orientation as a marketing perspective for primary healthcare organizations. *Int J Healthc Manag* 2023. https://doi.org/10.1080/20479700.2023.2248373.
7. Stocchi L, Ludwichowska G, Fuller R, Gregoric A. Customer-based brand equity for branded apps: a simple research framework. *J Mark Commun* 2021;27:534–63. https://doi.org/10.1080/13527266.2020.1752775.
8. What is brand awareness and why is it important? | Amazon Ads n.d. https://advertising.amazon.com/library/guides/brand-awareness (accessed January 7, 2024).
9. Healthcare branding: the complete guide in 2022 - Qualtrics AU n.d. https://www.qualtrics.com/au/experience-management/industry/healthcare-branding/ (accessed January 7, 2024).
10. Odoom R, Agyeman DO. Branding for small and medium sized healthcare institutions. *Heal Serv Mark Manag Africa* 2019:135–47. https://doi.org/10.4324/9780429400858-12/BRANDING-SMALL-MEDIUM-SIZED-HEALTHCARE-INSTITUTIONS-RAPHAEL-ODOOM-DOUGLAS-OPOKU-AGYEMAN.
11. Lowry PB, Vance A, Moody G, Beckman B, Read A. Explaining and predicting the impact of branding alliances and web site quality on initial consumer trust of e-commerce web sites. *J Manag Inf Syst* 2008;24:199–224. https://doi.org/10.2753/MIS0742-1222240408.
12. How marketing strategy impacts healthcare branding | Colormatics n.d. https://www.colormatics.com/article/healthcare-marketing-strategy/ (accessed January 7, 2024).
13. Mason AN, Narcum J, Mason K. Social media marketing gains importance after Covid-19. *Cogent Bus Manag* 2021;8. https://doi.org/10.1080/23311975.2020.1870797.
14. Kalhor R, Khosravizadeh O, Kiaei MZ, Shahsavari S, Badrlo M. Role of service quality, trust and loyalty in building patient-based brand equity: modeling for public hospitals. *Int J Healthc Manag* 2021;14:1389–96. https://doi.org/10.1080/20479700.2020.1762053.
15. Daugherty ML. Small business marketing strategies for physical therapy practice owners. *Int J Healthc Manag* 2021;14:710–6. https://doi.org/10.1080/20479700.2019.1692505.
16. Sillup GP, Dehshal MH, Namini MT. Pharmaceutical companies and physicians: assessing their relationship. *Int J Healthc Manag* 2013;6:276–80. https://doi.org/10.1179/2047971913Y.0000000049.
17. Brand awareness in healthcare: best practices n.d. https://socialclimb.com/blog/brand-awareness-in-healthcare-best-practices/ (accessed January 7, 2024).
18. Srivastava RK, Bodkhe J. Does brand equity play a role on doctors prescribing behavior in emerging markets? *Int J Healthc Manag* 2020;13 Sup1:1–11. https://doi.org/10.1080/20479700.2017.1409954.

Chapter 17

Introduction to Intellectual Property

17.1 Basics of Intellectual Property

Intellectual Property (IP) assets play a crucial role in various business transactions, notably within the healthcare sector [1]. This chapter offers a comprehensive overview of optimal strategies for safeguarding, transferring, and maintaining IP rights in the context of such transactions. IP pertains to the legal domain focused on safeguarding creations of the intellect, manifesting in diverse forms [2].

IP is a broad categorisation encompassing intangible assets owned and legally shielded by individuals or entities, barring external utilisation or implementation without explicit consent. An intangible asset, in this context, refers to a non-physical possession owned by a company or an individual [3]. The fundamental idea behind IP is that creations of the human intellect should be granted protective rights comparable to those safeguarding tangible assets, or physical property. Consequently, most developed economies have instituted legal frameworks to protect both tangible and intangible forms of property.

IP constitutes a broad category encompassing intangible assets, devoid of any physical presence, that emanate from human intellect. This diverse array of IP includes, but is not limited to, artworks, symbols, logos, brand names, and designs. In the context of today's knowledge-centric economy, corporations demonstrate a meticulous approach in identifying and securing IP due to its substantial and strategic value [2].

DOI: 10.4324/9781003475309-23

The creation of valuable IP necessitates significant investments in skilled labour, time, and intellectual resources. These investments underline the imperative for both organisations and individuals to rigorously safeguard their IP rights, thereby preventing any unauthorised access or use. This vigilant protection is particularly critical considering the potential risks associated with unauthorised exploitation, which could compromise the unique competitive edge conferred by IP.

The responsibility of extracting value from IP and preventing its unauthorised utilisation is central to the corporate agenda. Despite its intangible nature, IP often surpasses the monetary value of a company's physical assets, standing as a formidable source of competitive advantage. Consequently, companies ardently prioritise the protection and preservation of their IP holdings to maintain their strategic market position.

Numerous categories of IP are not explicitly delineated on the balance sheet as assets, primarily due to the absence of well-defined accounting principles for the valuation of each specific asset. However, the influence of such IP on the company's overall value is often discernible through its impact on the stock price, as market participants are well informed about its existence and potential implications for the company's worth.

Certain intangible assets, notably patents, are formally acknowledged as property within the accounting framework due to their inherent expiration dates. The quantification of these assets is achieved through the process of amortisation, an accounting methodology designed to systematically diminish the recorded value of an intangible asset over a predetermined period [4]. Amortisation serves the practical purpose of enabling companies to allocate a consistent amount annually for tax purposes, aligning with the gradual reduction in the asset's useful life.

For instance, consider a patent with a defined 20-year lifespan before entering the public domain. In this scenario, the company attributes a total value to the patent, and over the 20-year span, this value is evenly distributed through the amortisation process. Consequently, this annual amortisation serves to reduce the company's net income or profit for tax purposes. It is noteworthy that IP perceived to possess a perpetual life, such as trademarks, remains exempt from amortisation, as these assets lack a predetermined expiration date [2]. This nuanced understanding of the treatment of IP in financial accounting underscores the complexities involved in valuing and accounting for these intangible assets. There are several types of IP which are listed as follows.

17.2 Types of IP

Patents: A patent represents a proprietary entitlement granted to an inventor, typically sanctioned by a government agency, such as the US Patent and Trademark Office (USPTO). This legal document bestows upon the inventor exclusive rights to the particular invention, encompassing designs, processes, improvements, or tangible inventions like machines.

Prominent within the technology and software sectors, companies frequently secure patents to safeguard their innovative designs. As an illustrative instance, the patent for the personal computer, filed in 1980, stands as a notable example of this protective mechanism [2]. The application for this patent was initiated by Steve Jobs and three other colleagues at Apple (AAPL), underscoring the significance of patenting as a means for technology companies to assert exclusive rights over their groundbreaking innovations [2].

A patent constitutes an exclusive entitlement accorded to an invention, whether it be a product or a process. In essence, this legal protection is extended to innovations that present a novel approach or solution to a problem. In order to secure a patent, comprehensive technical details about the invention must be divulged to the public through a formal patent application. This disclosure requirement serves to disseminate knowledge about the invention, contributing to the collective pool of technical information while granting the inventor the exclusive right to exploit and control the use of the patented invention for a specified duration.

In essence, the patent owner wields a fundamental and exclusive authority to prevent or curtail any commercial exploitation of the patented invention by others. Put simply, the protective shield of a patent signifies that the patented creation cannot be manufactured, utilised, distributed, imported, or traded by external entities without obtaining explicit consent from the patent owner [5, 6]. It is crucial to underscore that the ambit of patent rights is circumscribed by territorial considerations, meaning that the exclusive privileges are confined to the geographic boundaries of the country or region where the patent has undergone formal filing and approval, adhering to the legal norms of that specific jurisdiction.

This protective jurisdictional framework is not perpetual; rather, it is conferred for a finite period, typically spanning 20 years from the date of submitting the patent application. This temporal limitation is instituted to strike a delicate balance between incentivising innovation by granting inventors exclusive rights and ensuring that these innovations eventually contribute to the public domain. The delineation of patent rights across territories, coupled with a defined duration, encapsulates a strategic approach to foster innovation, safeguard inventors' interests, and foster a dynamic exchange of ideas within the broader framework of IP law.

Copyrights: Copyrights confer upon authors and creators of original content the exclusive entitlement to utilise, replicate, or duplicate their intellectual creations. This legal protection extends to various forms of artistic expression, including literary works such as books and compositions by musical artists. Importantly, a copyright not only grants the original creators the authority to control the reproduction and dissemination of their work but also establishes the provision for these creators to authorise others through licensing agreements to utilise their creative output. Through this mechanism, copyright law aims to strike a balance between safeguarding the rights of content creators and fostering a regulated framework for the legitimate use and dissemination of creative works.

Copyrights and trademarks, often erroneously interchanged, serve distinct functions in IP law [7]. Copyrights, specifically, bestow exclusive authorship rights to creators of original works, encompassing publications, computer software, blog posts, website content, and photographs. Conversely, trademarks are utilised to safeguard brand identities. The imperative of copyright registration becomes evident when considering legal redress against plagiarism, allowing the use of the copyright symbol (©) even in the absence of formal registration. However, the pursuit of statutory damages and attorney fees necessitates the completion of the registration process with the USPTO. Copyright protections extend for the lifetime of the author plus 70 years, culminating in the entry of the work into the public domain [8].

Vigilance in establishing unambiguous ownership parameters is paramount, particularly in contractual agreements with contractors like website or software developers. Exercise caution when employing work-for-hire language, as joint copyright ownership entails equal rights and equitable profit-sharing among co-owners. Derivative works, which build upon preexisting material, represent a creative extension of copyrighted content, introducing complexities in discerning the original contribution and determining fair profit distribution. Navigating these intricacies demands a meticulous approach to ownership agreements and a nuanced understanding of copyright nuances to pre-empt and address potential challenges that may arise in such contractual scenarios.

Trademarks: A trademark serves as a distinctive symbol, phrase, or insignia that holds recognisability and serves to distinguish a product in the marketplace, legally setting it apart from other offerings [9]. This exclusive identifier is assigned solely to a particular company, indicating that the company holds proprietary ownership of the trademark, thereby precluding any unauthorised use or replication by others.

Frequently intertwined with a company's brand identity, a trademark encapsulates elements such as logos and brand names, imparting a unique visual and symbolic representation. As an illustration, the iconic logo and brand name of Coca-Cola are owned exclusively by the Coca-Cola Company, exemplifying the proprietary nature of trademarks and their pivotal role in establishing and protecting the distinct identity of a company and its products.

A trademark constitutes a distinctive sign with the capability to differentiate the goods or services of a specific enterprise from those offered by other entities. The safeguarding of trademarks is facilitated through IP rights, providing legal protection to prevent unauthorised use or imitation by competitors. At the national or regional level, the procurement of trademark protection necessitates the submission of a registration application to the pertinent national or regional trademark office, accompanied by the payment of requisite fees. On the global scale, two viable approaches are available: one may choose to individually submit a trademark application to the trademark office of each specific country where protection is desired, or alternatively, engage in the international registration system.

Fundamentally, the formal registration of a trademark establishes an unequivocal and exclusive right for its proprietor to employ the registered symbol. This exclusivity extends to the sole usage by the rightful owner or, alternatively, the granting of licenses to third parties in exchange for monetary compensation. The process of registration not only imparts legal certainty but also reinforces the position of the rights holder, particularly in the context of legal proceedings or prospective litigation scenarios. This fortified legal standing serves as a robust safeguard, solidifying the protective perimeter of the registered trademark and enhancing the proprietor's ability to assert and defend their exclusive rights in various commercial contexts.

The term of trademark registration typically spans a decade, with the provision for indefinite renewal contingent upon the payment of additional fees. These proprietary trademark rights, being inherently private, find enforcement through legal channels and the issuance of court orders. A trademark, in its diverse forms, may consist of a single word, a combination of words, letters, or numerals. Furthermore, the spectrum of trademarks extends to encompass visual elements such as drawings or symbols, three-dimensional attributes including the shape and packaging of goods, and non-visible signs like sounds or fragrances. This expansive array of possibilities underscores the limitless potential for distinct features to serve as identifiers in the commercial landscape.

In the contemporary healthcare landscape, the geographical scope of medical practices has expanded significantly, transitioning from localised operations to national or interstate establishments. Within this evolving paradigm, it becomes imperative for medical practices, excluding those merely identified by proprietor names, to secure trademark protection for their nomenclature. Consider the differentiation between generic names like "The Children's Hospital" and more distinctive alternatives such as "Komansky Children's Hospital" or "Morgan Stanley Children's Hospital," or the juxtaposition of "Urology Associates" with "Institute for Urologic Excellence." The act of trademarking not only establishes a unique brand for all products and services but also serves as a safeguard against consumer confusion, protects against counterfeit products, and acts as a deterrent to impostors.

The significance of trademarking transcends mere marketing considerations, emerging as a substantial business asset with the potential to generate revenue through licensing arrangements, serving as a pivotal element in franchising agreements, and proving instrumental in securing third-party financing. Particularly for startups, trademarks play a crucial role in fostering brand recognition, and through the designation of "intent to use," they can even be obtained for goods and services anticipated for future offerings. In the digital age, trademarks further facilitate online visibility, aiding patients in efficiently locating healthcare providers on the internet and social media platforms. The strategic acquisition and utilisation of trademarks, therefore, manifest not only as a marketing tool but as a multifaceted asset contributing to the overall success and sustainability of healthcare enterprises.

Franchise: A franchise represents a contractual arrangement where an entity, individual, or party, termed the franchisee, acquires a license to employ the name, trademark, proprietary knowledge, and operational methodologies of another entity known as the franchisor. Typically, the franchisee, often an entrepreneur or small business proprietor, assumes responsibility for managing the franchise or store. This licensing agreement affords the franchisee the privilege of marketing products or delivering services under the brand identity of the franchisor.

In reciprocation for this licensing privilege, the franchisor receives both an initial startup fee and ongoing licensing fees from the franchisee. This business model is exemplified by the practices of companies like United

Parcel Service (UPS) and McDonald's (MCD), which have effectively utilised franchising as a strategic approach to broaden their brand presence and extend their distribution networks. The franchise model, with its structured licensing framework, serves as a conduit for business expansion and collaboration between franchisors and franchisees.

Trade secrets: A trade secret denotes confidential information retained by a company, the value of which is contingent upon its non-disclosure. To safeguard IP maintained as a trade secret, it is prudent to establish non-disclosure agreements (NDAs) with contractors, manufacturers, distributors, and other relevant parties who may come into contact with such confidential information. Despite the enduring advantage of indefinite longevity, trade secrets carry inherent drawbacks, particularly the prospect of unauthorised discovery through reverse engineering or analogous processes.

The trajectory of trade secret litigation has consistently risen, with the healthcare sector notably witnessing a marked increase, particularly in domains such as computer technology, customer lists, proprietary pricing, and supplier relationships. Given these dynamics, a thorough and thoughtful analysis becomes imperative before opting to preserve IP as a trade secret. While the potential longevity of a trade secret is unlimited, a judicious evaluation is essential to assess the advantages against the concomitant risks and challenges associated with safeguarding sensitive information.

A trade secret constitutes a proprietary process or practice within a company that is not publicly disclosed, conferring an economic benefit or advantage upon the company or the possessor of the trade secret. The safeguarding of trade secrets demands active protection measures by the company, often stemming from extensive research and development (R&D) efforts. This is underscored by the common practice of employers requiring individuals to sign NDAs to reinforce the confidential nature of these proprietary elements.

Examples of trade secrets encompass a broad spectrum, ranging from designs, patterns, recipes, formulas, to proprietary processes. These trade secrets serve as foundational elements for crafting a unique business model that sets the company apart in the marketplace, thereby furnishing a competitive advantage. The deliberate protection of these confidential practices is essential for sustaining the distinctiveness of a company's offerings and ensuring a competitive edge in the industry.

The legal protection of business secrets is contingent on the legal framework, either being embedded within the broader concept of safeguarding against unfair competition or governed by specific provisions and case law addressing the protection of confidential information. While the determination of whether trade secret protection has been infringed upon relies on the particulars of each case, generally, unfair practices concerning confidential information encompass acts of industrial or commercial espionage, breaches of contract, and breaches of confidence.

It is noteworthy that a trade secret owner lacks the authority to prohibit others from utilising the same technical or commercial information if those parties independently acquire or develop such information through their own R&D, reverse engineering, marketing analysis, and similar means. Unlike patents, trade secrets, by virtue of not being publicly disclosed, do not offer "defensive" protection as prior art. For instance, if a specific process for producing Compound X is protected as a trade secret, another entity can secure a patent or utility model for the same invention if arrived at independently.

17.3 About Healthcare IP

Within the healthcare sector, the realm of IP law assumes a pivotal role as the impetus driving innovation. This force is instrumental in propelling progress in medication formulas, treatments, and medical technology, all geared towards the overarching objective of augmenting patient longevity and enhancing the quality of life.

Evidently, there is a discernible uptick in the acquisition of healthcare-related patents, a phenomenon particularly pronounced in an era where healthcare solutions are increasingly customised to individual needs. The surge is exemplified by the widespread adoption of wearable devices and health-centric applications among consumers. Empowered by smartwatches and fitness trackers, individuals now possess the autonomy to independently collect and monitor real-time health metrics.

The global popularity of wearable technology is a noteworthy trend, with market projections indicating a substantial surge from $24 billion in 2017 to a projected $70 billion by 2025. In this landscape, patents emerge as an indispensable tool for developers of wearable devices, serving as a robust mechanism to protect the foundational technologies that underlie their products. Beyond protective measures, patents play a pivotal role in sustaining the financial underpinnings of medical research and development undertaken by manufacturers. Acting as a shield, they empower manufacturers to safeguard their innovations from competitors, thereby ensuring a continuous trajectory of progress in the medical field.

The recent global upheaval stemming from the COVID-19 crisis further accentuates the paramount significance of IP rights within the healthcare domain. The escalating demand for innovative healthcare solutions mandates that manufacturers institute robust safeguards for their IP today. This proactive stance ensures an uninterrupted trajectory of development and enhancement in products and services, ultimately translating into benefits for patients on a global scale. Following are the ways by which the IP impacts the healthcare industry or entrepreneurship [10].

1. Connected medical devices have ushered in a transformative era in the methodology employed by physicians and researchers for the acquisition and analysis of patient data. These devices, operating in real-time, provide invaluable insights, particularly in the management of chronic diseases like asthma and sleep disorders. Additionally, they empower patients to capture clinically relevant information beyond the confines of a medical office, thereby augmenting the overall landscape of healthcare.

The advent of connected medical devices has given rise to novel approaches for medical professionals to comprehend patients' lives and behaviours. This wealth of information holds the potential to fundamentally reshape the delivery of healthcare services. Offering a myriad of options for physicians, patients, medical researchers, and health-conscious consumers, these devices signify a paradigm shift in the realm of personalised healthcare.

However, the continued evolution of connected medical devices is contingent upon the robust protection provided by patents. Patent protection stands as a linchpin, enabling developers and manufacturers to persist in innovation and exploration within the sphere of this technology. A well-strategised patent portfolio becomes instrumental in furnishing these entities with a sustained competitive advantage in the market [10].

In all stages of development, patents play a pivotal role in technology modernisation. Particularly for early-stage companies, patents serve as a yardstick by which investors assess the technological prowess of a company. Approved patents play a crucial role in securing funding through avenues such as venture capital or private equity investment. This influx of capital, in turn, facilitates further research and development initiatives, fostering a continuous cycle of innovation in healthcare-related products.

2. Wearable device manufacturers grapple with the intricate challenge of securing patents for the sophisticated technologies embedded within their products. To safeguard the multifaceted components of these devices, manufacturers integrate diverse technologies and seek patents for their composite solutions. Nevertheless, within the highly competitive landscape of the industry, the potential for overlapping patents surfaces, where the patent of one company may inadvertently infringe upon another's. In such scenarios, companies may explore the option of cross-licensing with competitors. This collaborative practice entails a reciprocal exchange of patents devoid of licensing fees, coupled with a mutual commitment to refrain from legal actions against one another. This collaborative approach facilitates both entities in pursuing research and development without hindrance, thereby fostering continuous innovations in healthcare-related technology.

Moreover, patents play a pivotal role in facilitating collaborations between traditional and digital device manufacturers. Through clearly defined patent rights for each party, the probability of ownership disputes is significantly mitigated. Companies engaged in joint ventures can establish agreements to license patent rights on terms mutually advantageous to them. This cooperative arrangement ensures the ongoing evolution of sophisticated and progressive healthcare-related technology.

3. Patents assume a pivotal role in affording medical device companies and developers the means to protect their proprietary technology, thereby facilitating a continuous trajectory of innovation and refinement in their product offerings. This significance is particularly pronounced in the healthcare sector, characterised historically by a gradual adoption of technological advancements.

In the current global health crisis, healthcare institutions confront a shortage of personnel, a challenge anticipated to escalate to 13 million by the year 2030. Access to advanced medical devices equips physicians and healthcare professionals to administer efficient healthcare services, mitigating challenges associated with antiquated systems. Consequently, this allows a dedicated focus on precise diagnostics and timely treatment administration, ultimately contributing to enhanced patient outcomes.

Wearable devices, serving as a notable example, emerge as potent tools for medical practitioners to monitor patients in real time and collect precise data, proving particularly instrumental in managing chronic diseases. Furthermore, this technological capability empowers patients to capture clinically relevant information beyond the confines of a traditional medical setting, fostering a more personalised and patient-centric approach to healthcare.

The preservation of IP rights through patents plays a pivotal role in propelling the development of innovative medical devices and technologies. This, in a cascading effect, contributes to the advancement of healthcare outcomes, ultimately benefiting patients through improved and personalised healthcare solutions.

4. Eminent corporations within the wearable device sector, including Samsung, Apple, Sony, and Philips, have accumulated a substantial repository of patents within their IP portfolios. These patents serve as robust safeguards, protecting the technological innovations that constitute the foundation of their products and facilitating a continuous trajectory of developing new and enhanced options for discerning consumers.

The increasing prevalence of wearable devices is attributable to their multifaceted benefits in the realm of health and wellness, encompassing the monitoring of disabilities and the detection of chronic diseases. These devices empower consumers to actively monitor their individual health metrics, thereby facilitating requisite lifestyle adjustments. Furthermore, wearable devices serve as conduits for the seamless exchange of health-related data with healthcare professionals, thereby contributing to the diagnosis and treatment of specific medical conditions.

The widespread accessibility of wearable devices on the mass market has democratically extended access to this technology, enabling a broader demographic to avail themselves of its advantages. This democratisation holds particular significance, as it endows individuals with the capability to assert control over their health, fostering the ability to make informed decisions pertaining to their overall well-being.

5. Safeguarding IP rights for devices like wearables in the healthcare sector not only serves the interests of the companies responsible for their creation but also upholds the fundamental rights of individuals. These rights encompass the right to life and the right to health, acknowledged in both national and international legal frameworks.

Moreover, the judicious application of technological innovations has the potential to enhance the accessibility and equity of healthcare services for underserved populations. Pioneers in the field are actively exploring avenues to render IP protection more financially viable in developing nations, thereby dismantling barriers that impede innovation in these regions. This endeavour holds promise for fostering inclusive advancements in healthcare technologies.

17.4 Infringement of IP

Connected to IP are specific entitlements recognised as IP rights. These rights are inviolable and cannot be encroached upon by individuals lacking authorisation. IP rights bestow upon owners the authority to prevent others from replicating, imitating, or exploiting their creations [11].

Infringement of patents occurs when an individual or entity utilises a legally protected patent without proper authorisation. Patents filed before June 8, 1995, retained a validity period of 17 years, whereas those filed after this date are valid for 20 years [12]. Following the expiration date, the details of the patent become publicly accessible [2].

Copyright violations transpire when an unauthorised entity replicates all or a portion of an original work, encompassing artistic creations, music, or literary works. It is pertinent to note that the duplicated content need not precisely replicate the original to qualify as an infringement. Trademark infringement takes place when an unauthorised entity employs a licensed trademark or a mark resembling the licensed trademark. For instance, a competitor may adopt a mark akin to a rival's to disrupt business and attract their customer base. Additionally, entities operating in disparate industries may adopt identical or similar marks in an attempt to capitalise on the robust brand images of other companies. This infringement across various IP realms necessitates vigilant legal enforcement to uphold the rights of the original creators and owners.

Infringement often occurs unintentionally. To mitigate the risk of legal action related to IP infringement, it is imperative to ensure that your business refrains from utilising copyrighted or trademarked material without proper authorisation. Additionally, exercise caution to ensure that your brand or logo does not bear excessive resemblance to those of others, to the extent that it could reasonably mislead individuals into associating it with another brand. Proactive measures in diligently assessing and differentiating IP assets can help forestall inadvertent infringements and safeguard against potential legal ramifications.

Conducting a thorough patent search is advisable to ascertain the originality of ideas. In the event that the ideas are not exclusive to you, exploring avenues for licensing through appropriate channels becomes crucial. Engaging the services of IP lawyers who specialise in this process ensures diligent examination to prevent the inadvertent use of protected intellectual assets belonging to others.

When enlisting individuals to perform creative work for your company, it is essential to ensure that the contractual agreement explicitly stipulates the transfer of ownership for any creative output generated. This provision safeguards against ambiguity and ensures that the IP becomes the unequivocal property of the company rather than the individual hired for the creative work. Clear contractual arrangements in this regard are pivotal for protecting the company's intellectual assets and avoiding potential disputes [13].

References

1. Schwartz HM. American hegemony: intellectual property rights, dollar centrality, and infrastructural power. *Rev Int Polit Econ* 2019;26:490–519. https://doi.org/10.1080/09692290.2019.1597754.
2. What is intellectual property, and what are some types? n.d. https://www.investopedia.com/terms/i/intellectualproperty.asp (accessed January 7, 2024).
3. Basant R, Srinivasan S. Intellectual property protection in India and implications for health innovation: emerging perspectives. *Innov Entrep Heal* 2016;3:57–68. https://doi.org/10.2147/IEH.S56236.
4. Li P. Health technologies and international intellectual property law: a precautionary approach, (1st ed.). 2013. Routledge. https://doi.org/10.4324/9780203550663
5. Ragavan S, Vanni A. (Eds.).Intellectual property law and access to medicines: TRIPS Agreement, health, and pharmaceuticals, (1st ed.) 2021. Routledge. https://doi.org/10.4324/9781003176602
6. Babyar J. Trade, intellectual property, and the public health bearing. *Heal Syst* 2023;12:123–32. https://doi.org/10.1080/20476965.2022.2062460.
7. What is intellectual property? n.d. https://www.wipo.int/about-ip/en/ (accessed January 7, 2024).
8. Intellectual property: what every medical practice needs to know n.d. https://www.medicaleconomics.com/view/intellectual-property-what-every-medical-practice-needs-to-know (accessed January 7, 2024).
9. Bhakuni H, Miotto L. Introduction: justice in global health. *Justice Glob Heal New Perspect Curr Issues* 2023:1–12. https://doi.org/10.4324/9781003399933-1/INTRODUCTION-HIMANI-BHAKUNI-LUCAS-MIOTTO.

10. How intellectual property law is impacting the healthcare industry - Lexology n.d. https://www.lexology.com/library/detail.aspx?g=080d5c38-8773-4b13-a434-710c618bb135 (accessed January 7, 2024).
11. What is intellectual property "theft" and how to avoid it? | Dennemeyer.com n.d. https://www.dennemeyer.com/ip-blog/news/what-is-intellectual-property-theft-and-how-to-avoid-it/ (accessed January 7, 2024).
12. Patent duration | LegalMatch n.d. https://www.legalmatch.com/law-library/article/patent-duration.html (accessed January 7, 2024).
13. Austin M. Licensing, selling and finance in the pharmaceutical and healthcare industries. *Licens Sell Financ Pharm Healthc Ind* 2016. https://doi.org/10.4324/9781315592343/LICENSING-SELLING-FINANCE-PHARMACEUTICAL-HEALTHCARE-INDUSTRIES-MARTIN-AUSTIN.

Chapter 18

Ethical and Legal Implications in the Healthcare Industry

18.1 Introduction to Ethical Issues in the Healthcare Industry

Whether one assumes the role of a healthcare administrator or a physician, the professional journey within the realm of healthcare unfolds as an experience marked by profound satisfaction and, concurrently, notable challenges. The healthcare field, characterised by its intricate nature, necessitates astute decision-making at every juncture [1, 2]. The choices made, encompassing the specific type of care a patient undergoes to the essential resource allocations for a health unit, inherently possess the potential to give rise to conflicts and complexities within the system [3].

The foremost and fundamental principle within the healthcare industry is encapsulated in the Latin maxim, "Primum non nocere," translating to "First, do no harm." While this directive may appear straightforward, the healthcare sector is characterised by its complexity, presenting all individuals in this demanding field with intricate choices on a daily basis. A comprehensive understanding of the diverse ethical dilemmas inherent in healthcare will provide the necessary empowerment to make judicious decisions that prioritise the well-being of patients. Unwavering adherence to elevated ethical standards constitutes an imperative for every healthcare professional, be they a doctor, nurse, or administrator [4].

A subfield of applied ethics known as "ethical concerns in healthcare" is concerned with the moral judgement that medical personnel must use when making decisions. The moral and ethical standards that underpin

DOI: 10.4324/9781003475309-24

the practice of medicine frequently show variances that are impacted by the unique features of every nation and its cultural milieu. Notably, across geographical and cultural boundaries, Tom L. Beauchamp and James F. Childress have developed a uniform approach to ethical dilemmas in healthcare [4]. These recommendations are an invaluable tool for medical workers, providing a framework for resolving moral conundrums that crop up when medical ethics are called into question.

The ethical standards framework embraced within the healthcare sector encompasses four pivotal principles that guide ethical decision-making [4]:

- Autonomy—Prioritise eliciting and respecting the patient's wishes to safeguard their autonomy.
- Beneficence—Strive to act in the best interest of the patient, ensuring actions are geared towards their well-being.
- Justice—Engage in due process to ascertain the boundaries of healthcare, ensuring fair and equitable distribution of resources and treatments.
- Non-Maleficence—Address and mitigate harm, meticulously determining strategies to prevent its occurrence.

Effective collaboration between healthcare practitioners and patients is essential for understanding and reconciling the diverse needs and desires inherent in human beings. An ethical quandary emerges when considering whether it is appropriate for health practitioners to administer a blood transfusion to a Jehovah's Witness, even when it could potentially save their life, given the conflict with the patient's religious beliefs.

In the modern healthcare landscape, a collaborative and communicative approach with patients is paramount to prevent ethical transgressions. Medical professionals must refrain from making assumptions about a patient's preferences and requirements. The ethical obligation involves seeking explicit informed consent through meaningful dialogue and adhering to established safety protocols, thereby ensuring the upholding of ethical principles in healthcare delivery.

There are always going to be a lot of medical ethical questions while a patient is receiving treatment. Problems pertaining to waiting lists, fair access to healthcare resources, and choosing the best course of action are all considered ethical issues. It is imperative to acknowledge that ethical concerns deviate from legal limitations; an action may be considered immoral even when it complies with all applicable laws. The distinction between ethical and legal dimensions underscores the complexity and nuance inherent in navigating ethical quandaries within the realm of healthcare [5].

An illustrative scenario is the perpetual backlog in the emergency room; from a legal standpoint, healthcare facilities or staff are not obligated to expedite their work. However, ethically, it may be considered appropriate to bring the matter to the attention of the hospital's management team, exploring avenues to enhance efficiency and accommodate more patients promptly.

Navigating ethical considerations in healthcare is a complex undertaking, often entailing situations where the optimal decision may not align seamlessly with what feels ethically sound. The challenge intensifies, particularly in life-threatening emergencies, where ethical dilemmas demand swift and thoughtful resolution. The following outlines common examples of such dilemmas and elucidates best practices for healthcare management, professional organisations, and healthcare practitioners (HCPs) to uphold ethical behaviour in these demanding situations.

18.1.1 Major Ethical Issues

The landscape of the medical world is in a perpetual state of evolution, driven by continuous advancements in medical technology. In response to these changes, physicians find themselves grappling with an array of novel ethical issues. According to a recent Sermo survey gauging the concerns of physicians [3], the most prevalent ethical challenge reported is the delicate balance between ensuring the quality of patient care and optimising efficiency, reflecting the evolving nature of healthcare delivery.

The survey delves further into the various ethical conundrums that face doctors. Respondents highlighted important issues like managing the distribution of scarce donor organs, addressing end-of-life concerns, ensuring access to care, protecting doctor–patient confidentiality, and allocating limited medications or tools of support. These findings underscore the intricate ethical landscape that medical professionals navigate as they strive to provide optimal patient care amid the evolving dynamics of the healthcare industry.

Resolving an ethical dilemma in the field of medicine is seldom a straightforward process, unlike the binary nature of lawful and unlawful actions. The complexities inherent in medical ethics demand careful examination and nuanced considerations. Let's delve into some of the pivotal medical ethical issues confronting healthcare providers and staff in the contemporary landscape [3, 5].

18.1.1.1 Do-Not-Resuscitate (DNR) Orders

The ethical implications surrounding do-not-resuscitate (DNR) orders have consistently been a matter of debate in the medical field, as evidenced by the 17% of survey participants citing end-of-life issues as a noteworthy ethical concern. A DNR order explicitly directs healthcare providers against administering cardiopulmonary resuscitation (CPR) in instances of cardiac or respiratory arrest, and its issuance requires consultation with both a physician and the patient. However, ethical complexities arise when a patient unequivocally displays a DNR order, as illustrated by a 2017 case wherein a Florida man had the directive tattooed onto his chest. Despite scepticism regarding the gravity of the tattoo, the intensive care unit (ICU) opted to respect the message, refraining from initiating CPR. Such situations exemplify the profound ethical and moral challenges inherent in end-of-life decision-making.

From an ethical standpoint, the issuance and interpretation of DNR orders demand a delicate equilibrium between mitigating potential treatment-related pain and assessing it against anticipated benefits. Consequently, addressing end-of-life issues emerges as a formidable challenge for medical professionals and healthcare organisations, requiring a nuanced approach to ethical decision-making.

18.1.1.2 Patient Confidentiality

With 15% of respondents in a recent survey citing doctor–patient secrecy as their top ethical concern in the practice of medicine, patient privacy and confidentiality emerge as critical legal and ethical considerations in healthcare. The security and privacy of all patient medical records is a critical responsibility that healthcare professionals must fulfil. This obligation is reinforced by the Health Insurance Portability and Accountability Act (HIPAA), which states that infractions may result in suspension or termination.

However, the landscape surrounding patient privacy encompasses nuanced areas that may inadvertently give rise to ethical quandaries in healthcare. Hence, it is imperative to approach these ethical dilemmas with a foundation of empirical knowledge and steadfast ethical principles rather than circumventing them. An illustrative scenario involves contemplating whether withholding certain aspects of a patient's medical condition could potentially cause more harm than good. Resolving such ethical issues demands a nuanced consideration that takes into account varied perspectives shaped by personal beliefs and professional life experiences. This underscores the necessity for a principled and informed approach to ethical decision-making within the healthcare domain.

18.1.1.3 Mandatory COVID-19 Vaccinations

The benefits and drawbacks of enacting mandatory COVID-19 vaccination laws have been the subject of intense debate among international public health authorities. This is an important issue that has an impact on healthcare providers as well as the healthcare sector as a whole. Numerous countries, both in Europe and Asia, have enacted mandatory vaccination requirements for all personnel within the healthcare sector.

This contentious issue has become one of the most fervently debated current ethical dilemmas within healthcare organisations, prompting profound questions about the intersection of individual medical liberty and the broader societal needs. The ethical considerations delve into the complexities of compelling individuals to undergo treatment against their will. Factors such as healthcare costs, personal and religious beliefs, patient outcomes, and long-term consequences contribute to the multifaceted nature of this debate, demanding a careful examination of the ethical implications inherent in mandating medical interventions.

18.1.1.4 Accessibility of Healthcare

This ethical quandary extends to questions about the propriety of refusing medical treatment to individuals based on their insurance status and the ethics surrounding hospitals charging exorbitant amounts for treatment. Beyond being an ethical concern within the field of medicine, this issue permeates political discussions, reflecting broader societal considerations about the operation of healthcare systems. The ongoing debate, spanning decades in the United States, persists with no clear resolution in sight. Politically, this is perceived as the defining ethical issue of our time, with the juxtaposition of profitability versus public health presenting a direct and enduring conflict.

18.1.1.5 Allocating Limited Donor Organs, Medication, and Tools of Support

According to the findings from the Sermo survey [3], more than 10% of respondents highlighted the allocation of limited donor organs as the primary ethical concern in healthcare, while an additional 17% expressed a similar viewpoint regarding the distribution of limited medication and tools. Remarkably, in the United States, more than 95% of people declare that they would be prepared to donate their organs, yet only 58% of them have official donor registrations. The data also show that, although fewer than 4% of the population is waiting for an organ transplant, the bulk of organs in the United States come from deceased donors, with a much smaller proportion coming from living donors.

This data underscores the pressing need for comprehensive research on the ethical considerations surrounding the allocation of limited donor organs, especially in the context of enhancing the number of living donors. An exploration of the ethical dimensions of this issue is pivotal in informing policies and strategies that can effectively address the existing disparities in organ availability, contributing to the augmentation of the pool of living donors.

Furthermore, managing medication and tools of support poses challenging ethical considerations for contemporary healthcare practitioners within the health system. Medication, being a primary method for curing diseases and serving as adjunctive therapy in conjunction with other treatments such as surgery, are indispensable. However, the limited supply of some medication introduces a moral quandary, where not all patients receive the necessary medication, leading to ethical conflicts.

A parallel challenge arises with the tools of support available to physicians, varying significantly across different regions. Many rural areas, unfortunately, grapple with a shortage of modern medical tools, compounding the ethical considerations within healthcare. However, the concept of support extends beyond the physical tools and encompasses multiple dimensions, often making it intricate to identify ethical challenges arising from supply shortages.

The demanding nature of a doctor's profession, especially when on-call at the hospital, leaves little time for rest between shifts. To effectively serve as patient advocates, doctors require support from colleagues and loved ones, crucial for navigating the challenges inherent in their work. Striking a balance between assisting patients and not neglecting personal relationships and hobbies is imperative. Significantly, when survey participants were asked about their sources of support for addressing ethical dilemmas and ethical principles, 31% mentioned colleagues, while 30% sought guidance from professional superiors.

18.1.2 Handling Ethical Issues

The presence of moral dilemmas and ethical issues imposes a significant burden on nurses, doctors, and administrative staff within the healthcare profession. While the primary focus should be on preventive measures, addressing ethical missteps becomes an integral aspect of the professional landscape. This necessity arises from the fact that a substantial 47% of physicians have encountered ethical dilemmas in the course of their own practices.

In a recent survey, healthcare professionals responded to ethical violations in various ways. Approximately 36% opted to report the conduct to relevant clinical authorities, such as the hospital's peer review body or the local/state medical society. Another 22% prioritised protecting patient privacy, while 17% chose to report the suspected violation to appropriate authorities. Additionally, 14% directly confronted individuals involved in the ethical breach, and 9% reported the conduct to the state licensing board. These diverse responses underscore the intricate nature of ethical decision-making within the healthcare domain, reflecting the nuanced approaches adopted by professionals when confronted with ethical challenges.

Addressing ethical violations at a practice involves a systematic approach. Initially, it is crucial to acknowledge and investigate potential ethical issues. Fostering an environment that encourages open communication across various disciplines, irrespective of seniority or primary responsibility, is essential in this process.

Following identification, the ethical violation should be categorised as either a process issue or a regulatory issue. It is important to recognise that actions can be deemed unethical even if they align with legal standards. For instance, when faced with a parent refusing to support a patient's wishes, it becomes imperative to refer to the clinic or hospital's policy to determine the appropriate course of action.

Real-life examples of ethical dilemmas in healthcare include cases involving informed decision-making with multiple family members, instances of sexual harassment, ethical considerations in managing electronic health records, and conflicts arising from recommendations that contradict religious or personal beliefs with business ethics. Managing these common ethical issues is crucial, as they carry potential consequences and necessitate careful consideration with the well-being of patients and employees at the forefront.

Addressing ethical dilemmas in healthcare may necessitate updates to internal rules and work processes. This involves consulting frontline staff members, the administrative body, and the legal team to ensure a comprehensive and well-informed approach. To effectively handle ethical challenges in medicine, adopting the following strategies is recommended:

- Provide Regular Refreshers on Ethical Codes: Regularly refresh healthcare professionals on relevant ethical codes, such as American Speech-Language-Hearing Association (ASHA) Code of Ethics, and reinforce best practices in data privacy.
- Break Silos for Collaborative Discussion: Encourage open communication and collaboration by breaking down silos within the organisation. Facilitate discussions on ethical issues, involving different staff members with diverse perspectives.
- Establish Clear Chain of Command: Define a clear chain of command for addressing ethical issues, specifying when escalation is necessary or when group decision-making processes should be activated.
- Prioritise Patient Wishes: Encourage the patient's wants to be given priority in the setting. This entails placing a strong emphasis on informed consent, closely examining dubious prescription orders, and, when necessary, consulting family members.
- Involve Committees or Experts in Ethics: Think about working with medical teams in conjunction with ethics experts. As an alternative, create ethics committees to ensure a comprehensive ethical appraisal by methodically reviewing the rights of patients, decision-making procedures, and moral standards in certain situations.

18.2 Legal Issues in the Healthcare Industry

The healthcare industry is in a state of continual evolution, requiring not only the governance, risk, and compliance (GRC) function but also for providers and support staff to stay abreast of the evolving legal landscape. In health systems, understanding the changing legal realities is imperative as lawmakers, payers, patients, and various stakeholders adapt to new dynamics [6–8]. The ongoing complexities posed by the COVID-19 pandemic have significantly influenced laws pertaining to healthcare. It is crucial for providers and administrators to be cognisant of seven key legal issues that have emerged, shaping the current healthcare landscape [9].

18.2.1 Telehealth Law

The year 2020 witnessed a significant surge in telehealth law, with this rapidly expanding legal domain experiencing unprecedented growth. Notably, various measures were implemented, such as waivers to reduce telehealth payment barriers, safeguards for patient protection, and audits targeting fraud reduction. In 2021, the trajectory suggests a continued expansion of telehealth coverage, necessitating an updated understanding of the relevant regulations.

Providers are urged to remain informed, particularly concerning the Centers for Medicare & Medicaid Services (CMS) List of Medicare Telehealth Services. It is essential for billing staff to be well-versed in both permanent and temporary codes applicable to reporting telehealth services. Furthermore, a comprehensive grasp of the multitude of regulations instituted in the past year is imperative.

Despite efforts to reduce barriers to telehealth, it is noteworthy that the Office of the Inspector General (OIG) has concurrently increased its audits in this domain. Telehealth providers are advised to adopt a proactive approach, conducting thorough reviews of billed claims and assessing the compliance of their telehealth programmes to ensure alignment with federal requirements.

18.2.2 PHI and HIPAA Compliance

The Health Insurance Portability and Accountability Act (HIPAA) is poised for substantial updates, as outlined in the proposal unveiled by the Office for Civil Rights (OCR) in December 2020. The primary focus of these proposed amendments is to enhance a patient's entitlement to access protected health information (PHI) while concurrently reducing obstacles in healthcare operations and value-based reimbursement systems.

Additionally, the plan recommends replacing the "professional judgement" criterion for PHI disclosures and uses with a standard based on a sincere belief in the person's best interests. Additionally, it suggests extending the permission to reveal PHI in cases where a harm is judged to be "severe and reasonably foreseeable" in order to avert a threat to health or safety. Additionally, the proposal eliminates the requirement to obtain a patient's written acknowledgment of a provider's Notice of Privacy Practices (NPP) and modifies the content requisites of the NPP.

Regarding HIPAA compliance, one significant development for 2021 is the expected rise in OCR enforcement proceedings through the HIPAA Right of Access Initiative. Since 2019, this initiative—which was implemented to protect people's right to timely access to their health records—has given rise to 18 enforcement proceedings and settlements totalling $3,500 to $160,000. All providers, no matter how big or small, should carefully review the processes for answering requests for patient records and make sure that they happen on time. According to the Department of Health & Human Services (HHS), the current standard is that access to requested information must be granted within 30 days, and in the event that access is denied, a written explanation must be supplied. The request must be completed within 60 days, even in cases where there are legitimate delays; this gives only single extension for everyone.

18.2.3 Employers' Responsibilities in Healthcare and Maintaining Safe Working Conditions

In 2021, healthcare employers faced heightened scrutiny and potential liability for their employees' exposure to and contraction of COVID-19, along with labour-related issues stemming from the pandemic. The notable increase in lawsuits, including class action suits, filed by employees alleging violations of federal and state regulations related to employee safety and labour matters underscored the dynamic legal landscape surrounding these issues.

Ensuring a safe working environment for healthcare workers became paramount, with guidance from the Centers for Disease Control (CDC) and the Occupational Safety and Health Administration (OSHA) serving as crucial references for healthcare provider organisations. OSHA emphasised the importance of developing and implementing infection control and preparedness plans, effectively communicating these plans to workers through comprehensive training, and conducting risk assessments while adhering to the hierarchy of controls for worker protection.

A significant percentage of COVID-19–related cases filed by employees were labour-related issues, namely allegations of retribution, unlawful termination, or improper denial of leave. Employers in the healthcare industry had to negotiate a complicated regulatory environment that included state and federal laws governing labour practises. It was important to note two government acts that addressed the pandemic:

- First Families to Respond to the Coronavirus (FFCRA): Mandated that, for COVID-19–related reasons, firms with fewer than 500 employees offer job-protected leave. This included situations in which workers had to take care of a young child or had a medical emergency. Employers understood that certain employees were excluded from certain leave benefits, particularly when it came to healthcare workers.
- Worker Adjustment and Retraining Notification Act (WARN): Employers with 100 or more workers are required to notify workers in advance of any permanent job location closures or mass layoffs. A recent court decision removed an exemption to the notice requirement by making it clear that COVID-19 would not be regarded as a "unforeseeable business condition."

Healthcare employers were urged to proactively address these challenges, staying abreast of evolving regulations and implementing robust measures to safeguard both employee well-being and legal compliance.

18.2.4 Nurseries and Long-Term Care Facilities

The COVID-19 epidemic posed serious issues for long-term care (LTC) facilities, nursing homes, and skilled nursing facilities. As a result, new government guidelines and regulations were put into place to guarantee the standard of care provided by these organisations. The CMS began conducting targeted inspections of most nursing facilities in March 2020. These inspections continued throughout 2021, necessitating that providers stay informed about evolving guidance from CMS. Similar to telehealth providers, LTC and nursing facilities were advised to reference the CMS COVID-19 Emergency Declaration Blanket Waivers for Health Care Providers to navigate the changing regulatory landscape. Several blanket waivers, excluding certain requirements for nursing homes, expired, leading to the enforcement of these requirements.

An additional significant advancement was the mandate by CDC for LTC and nursing homes to submit weekly reports on COVID-19 data. These reports had to cover suspected and confirmed cases among residents and staff, overall deaths, deaths among residents and staff related to COVID-19, availability of supplies for personal protective equipment (PPE), ventilator capacity in the facility, resident beds and census, resident access to COVID-19 testing, and staffing shortages.

18.2.5 The False Claims Act

The False Claims Act (FCA) resulted in settlements and judgements totalling over $2.2 billion for the Department of Justice (DOJ) in 2020, with the healthcare sector accounting for $1.8 billion of this sum. The DOJ was able to correct erroneous claims for federal funding because of the FCA, which was an important tool in the fight against healthcare fraud. The 2020 actions covered a wide range of healthcare-related firms, including hospitals, pharmacies, hospice organisations, doctors, hospitals, producers of drugs and medical devices, managed care providers, and laboratories. During this time, a number of tendencies in FCA settlements appeared, and it is expected that these would continue:

- A Rise in Lawsuits Filed by Whistleblowers: Qui tam lawsuits, initiated by whistleblowers with critical inside information crucial for identifying potential fraud, accounted for $1.6 billion of the FCA cases in 2020. As a reward for their contributions, whistleblowers shared in the money recovered by the DOJ, with the government disbursing $309 million in whistleblower payouts.
- Increased Personal Accountability: Notably, there was an upswing in settlements holding individuals accountable. In some instances, individual doctors from medical practices agreed to significant payments (e.g., $4.25 million) to resolve civil allegations related to illegal kickbacks.
- Biggest Drug Manufacturer Recoveries Affect Medicare Patient Copayments: The most substantial recoveries involved pharmaceutical manufacturers that funded co-payments for Medicare patients to safeguard elevated drug prices. Two manufacturers each paid over $148 million to settle the claims of illegally covering patient copays for their drugs.
- Typical Fraud Schemes: The prevalent fraud schemes identified in 2020 included opioid-related fraud, followed closely by kickback schemes.

These trends underscored the DOJ's continued commitment to combating healthcare fraud through legal avenues, emphasising the importance of whistleblowers and holding individuals accountable for fraudulent practices.

18.2.6 Patient Safety and Healthcare Inequity

In 2021, racial and ethnic disparities in healthcare emerged as a prominent patient safety concern, highlighted by discrepancies in medical care access, testing, and vaccination among minorities during the COVID-19 pandemic. Notable studies underscored the extent of the problem:

- According to the CDC, 32.5% of COVID-19 deaths were among the Hispanic or Latinx community, although comprising only 18.5% of the country's total population.
- In six of the eleven patient safety quality measures, black patients received care that was noticeably worse than that of white patients, according to a study done by the Urban Institute's Health Policy Center. These metrics, which included five of the seven safety metrics pertaining to surgery, quantified the frequency of adverse patient safety events.

In response to these disparities, healthcare organisations were urged to allocate resources to enhance health equity by undertaking the following measures:

- Incorporated Health Equity into Organisational Strategy: Integrated health equity considerations into the overall strategy of the organisation and educated employees on the importance of addressing disparities.
- Assessed Organisational Culture: Evaluated the organisational culture concerning health equity and established goals to rectify any identified weaknesses in promoting equitable healthcare practices.
- Engaged with Community Resources: Collaborated with community resources and actively participated in their initiatives aimed at improving health outcomes and accessibility for minority populations.
- Addressed Racism within the Organisation: Confronted and addressed any instances of racism within the organisation, fostering a culture of inclusivity and diversity. Developed a cultural competence strategy to ensure equitable treatment for all individuals.

By taking these proactive steps, healthcare organisations contributed to the broader effort of mitigating racial and ethnic disparities in healthcare and fostering a more inclusive and equitable healthcare system.

18.2.7 Universal Availability of Healthcare

Although getting healthcare has always been difficult for many, a new CDC survey found that four out of ten adults in the United States put off getting care because of COVID-19–related problems. Furthermore, throughout the pandemic, 32% of adults skipped routine treatment and 12% of adults decided not to seek emergency care. According to the report, some groups are more negatively impacted by these difficulties than others on racial grounds and those with chronic illnesses, and unpaid family caregivers.

One of the most important components of the socioeconomic determinants of health was access to healthcare. It's possible that things like poor transportation and insufficient healthcare resources contributed to barriers to accessing treatment, but the biggest obstacle was still not having insurance. A greater proportion of Americans now have insurance as a part of the Affordable Care Act. The Act's Medicaid Expansion provisions were particularly important in lowering health coverage gaps and enhancing patient access.

References

1. Pakkanen P, Häggman-Laitila A, Kangasniemi M. Ethical issues identified in nurses′ interprofessional collaboration in clinical practice: A meta-synthesis. *J Interprof Care* 2022;36:725–34. https://doi.org/10.1080/13561820.2021.1892612.
2. Midkiff DM, Joseph Wyatt W. Ethical issues in the provision of online mental health services (Etherapy). *J Technol Hum Serv* 2008;26:310–32. https://doi.org/10.1080/15228830802096994.
3. Ethical issues in healthcare & medical challenges | Sermo n.d. https://www.sermo.com/resources/ethical-issues-in-healthcare/ (accessed January 8, 2024).
4. Big thinkers: Thomas Beauchamp & James Childress - the ethics Centre n.d. https://ethics.org.au/big-thinkers-thomas-beauchamp-james-childress/ (accessed January 9, 2024).
5. Current ethical issues in healthcare - Florida tech online n.d. https://www.floridatechonline.com/blog/healthcare-management/current-ethical-issues-in-healthcare/ (accessed January 8, 2024).
6. Clark A, Prosser J, Wiles R. Ethical issues in image-based research. *Arts Health* 2010;2:81–93. https://doi.org/10.1080/17533010903495298.
7. Bressler MY, Girard A, Felice S, Kiehlmeier D, Blazey W, Zampella JG. Ethical and legal issues for medical professionals using social media. *J Leg Med* 2021;41 sup1:5–8. https://doi.org/10.1080/01947648.2021.1914471.
8. Karcher NR, Presser NR. Ethical and legal issues addressing the use of mobile health (mHealth) as an adjunct to psychotherapy. *Ethics Behav* 2018;28:1–22. https://doi.org/10.1080/10508422.2016.1229187.
9. 7 Current legal issues in healthcare | symplr n.d. https://www.symplr.com/blog/7-current-legal-issues-in-healthcare (accessed January 8, 2024).

Chapter 19

Launching a Startup: Rules and Steps

19.1 Things to Know before Launching a Startup

Aspiring healthcare entrepreneurs may feel that creating a medical firm is a relatively simple endeavour with high prospects of success [1]. Healthtech is one of the most well-funded and fastest expanding sectors. According to research, the digital health business will be valued $551.09 billion by 2027, up from its present value of $220.16 billion. Funding for health technology startups in Q3 2021 exceeded Q2 2020 levels, with 72% of the funds going to seed and early-stage companies. One of the most well-funded and quickly expanding businesses, digital healthcare, is predicted to reach $42.22 billion by 2027 (Figure 19.1). Experts predict that by the end of 2022, the digital fitness and well-being category will be able to capture a significant portion of this total market revenue, with a total revenue of over $17.71 billion.

The reality is that commencing a startup in the healthcare industry is particularly challenging. Challenges arise from stringent regulations, elevated safety standards, and intricate health relations, posing obstacles for numerous companies aspiring to introduce novel products into the market [3]. Achieving success in this domain requires a profound understanding of launching a healthcare startup within the current regulatory environment.

Prominent medical institutions may not consistently embrace cutting-edge technology. According to the 11th Annual Report by InstaMed, a substantial

DOI: 10.4324/9781003475309-25

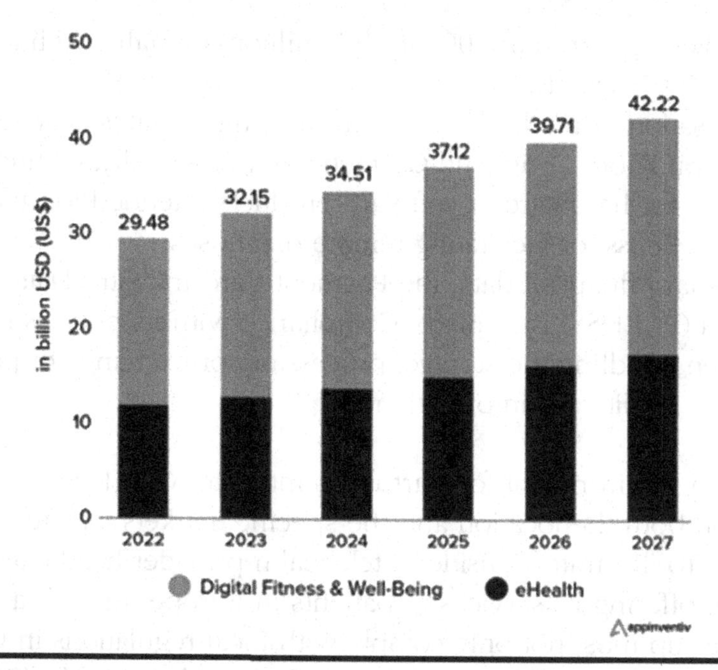

Figure 19.1 Growth of the digital health market over the years [2].

84% of healthcare providers continue to favour paper-based manual procedures for payment and patient data collection, notwithstanding the associated challenges [2]. Additionally, there exists a collective preference among doctors, nurses, and patients for software products that demonstrate user-friendly attributes.

Healthcare providers, apprehensive that the integration of new solutions could further complicate their already intricate workflows, exhibit reluctance towards adopting innovation. Moreover, a substantial 64% of physicians perceive technological complexity as the primary impediment to the acceptance of remote solutions among patients [2]. Lastly, medical professionals require substantiated evidence demonstrating that digital technologies enhance the quality of medical care. Consequently, the industry anticipates startups to substantiate their concepts with research founded on empirical evidence.

Aside from the aforementioned, several restrictions are also required, as covered in earlier chapters. Medical businesses have hundreds of federal and local laws to juggle. The primary US healthcare rules are as follows:

1. The Health Insurance Portability and Accountability Act (HIPAA) governs how personal health information is protected (medical data you can use to identify a person). The act describes the security, privacy, and electronic medical data interchange standards. Penalties for HIPAA

violations range from $50,000 to $1.5 million annually, with a maximum of $50,000 per event.
2. The utilisation of medical devices in the United States is governed by the Federal Food, Drug, and Cosmetic Act. Essentially, certification must be sought for hardware or software products intended for managing health, wellness, or facilitating remote diagnoses.
3. To safeguard financial data, the Payment Card Industry Data Security Standard (PCI DSS) is in place. Compliance with its policies is imperative when handling the storage, processing, or exchange of patients' bank and credit card information.

The regulatory environment for startups is intricate, with legal obligations contingent on both the location and the specific markets targeted by the company [4]. To illustrate, consider a telehealth provider headquartered in California, offering its services to patients in Europe. In such a scenario, the startup must not only comply with local regulations in California but must also adhere to the stringent provisions outlined in the General Data Protection Regulation (GDPR) and the UK General Data Protection Regulation (UK-GDPR).

GDPR, applicable across European Union member states, establishes comprehensive standards for data protection and privacy. Compliance with GDPR entails implementing robust measures for collecting, processing, and storing personal data, especially crucial in the healthcare sector where sensitive patient information is involved. Similarly, the UK-GDPR, governing data protection in the United Kingdom, adds an additional layer of compliance for businesses operating in or serving the UK market.

By following this complex regulatory terrain, foreign companies need to follow the local laws, irrespective of the country. For example, a California-based telehealth provider, which serves patient in the Europe, should not only ensure to adhere to local laws but also demonstrate a commitment to upholding rigorous data protection standards, thereby fostering trust and credibility among its patient base in Europe. This nuanced approach to regulatory compliance underscores the significance of a comprehensive understanding of legal frameworks in different regions to mitigate risks and uphold ethical business practices.

Entrepreneurs embarking on the journey of launching a healthcare startup must not underestimate the intricacies associated with health data [5]. Presently, 36% of medical record administrations encounter challenges in efficiently exchanging patient information. Consequently, to integrate your data into the existing system seamlessly, it is imperative that your product demonstrates compatibility with the array of solutions employed by medical organisations. This includes not only interfacing with diverse medical devices but also accommodating various data formats and terminologies utilised within the healthcare sector.

Launching a healthcare startup requires addressing the complexity of health data, given that 36% of medical record administrations encounter difficulties in sharing patient information. To seamlessly integrate your product into existing systems, it must comply with interoperability standards like Fast Healthcare Interoperability Resources (FHIR), Electronic Data Interchange (EDI), and Health Level Seven (HL7). These standards ensure compatibility with diverse healthcare solutions, devices, and data formats. Adhering to such standards becomes crucial for effective communication and data exchange within the healthcare industry.

In the conservative healthcare industry, trust in fledgling startups is often lacking. Consequently, building connections with prominent hospital chains, insurance companies, and healthtech firms becomes imperative for progress. A well-structured business plan, inclusive of a roadmap, illustrates the scalability of the startup and provides assurance that it won't dissipate after securing initial funds. The presence of respected entrepreneurs and medical professionals on the board enhances credibility. Furthermore, substantiating the validity, compliance, and safety of the product is pivotal in gaining trust within the industry.

Penetrating the healthcare sector demands a pragmatic mindset, given its resistance to swift upheavals. Excluding wearables, there has been a scarcity of digital products achieving viral status in the past ten years. The industry's entrenched bureaucracy contributes to protracted sales cycles. Prudent financial outlooks are imperative for any groundbreaking healthcare startup, necessitating a readiness to sustain the venture for at least five years before realising profitability. This underscores the significance of a sturdy business strategy and assembling an adept team for sustained success.

19.2 Major Reasons for Failure of Healthcare Startups

The appetite for groundbreaking healthcare technology and solutions is substantial. The pandemic, for instance, has triggered a noteworthy surge in the embrace of telehealth and telemedicine solutions by both patients and medical institutions. Surveys indicate that more than 70% of the US

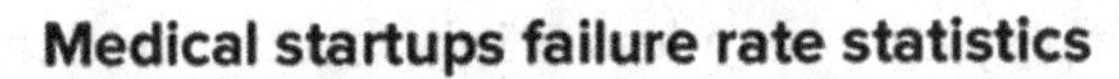

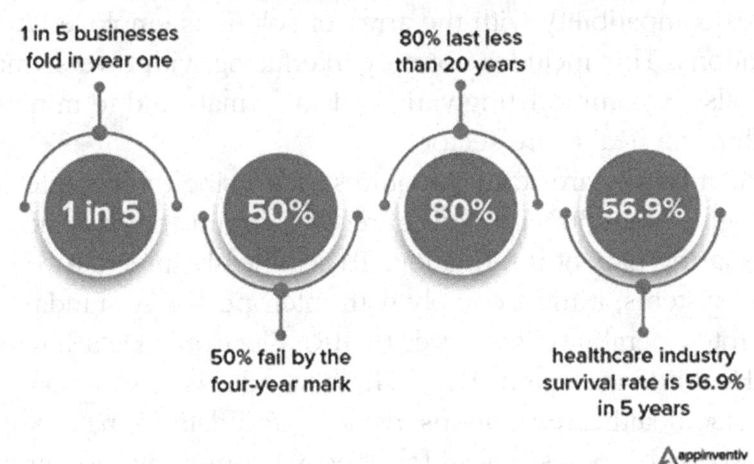

Figure 19.2 Startup failure rates in the healthcare domain [2].

population expresses a willingness to engage with telemedicine, while approximately 76% of healthcare businesses are prepared to incorporate telemedicine coverage into their insurance plans. Statistics show that for the healthcare domain (Figure 19.2), one in five businesses fold in a year. In addition to this, 50% are reported to fail by the fourth year followed by 80% by 20 years. Also, the healthcare industry has showed a low survival rate of approximately 57%.

Therefore, why can't new healthcare startup firms prosper if they're providing the proper solutions? Let's determine the main causes of their failure [2].

19.2.1 Failure to Adhere to Industry Norms

Starting a healthcare business necessitates extensive preparation and early legal advice. The difficulties faced by owners in managing licencing, regulatory compliance, testing, industry bureaucracy, and accreditation often result in the failure of medical companies. On the other hand, medical business owners that carefully carry out research increase their chances of success.

Before starting, it's important to take into account a number of factors: it's crucial to pick service compliance carefully because there are differences in the legal requirements and approvals for healthcare technology firms. Some healthcare business ideas may be quickly brought to market because of their shorter preparation cycles. One such example is "software-only medical goods." On the other hand, certain medical startups include complex legal procedures, such ventures governed by US Food and Drug Administration (FDA) regulations and using clinical items of a professional calibre.

19.2.2 Ineffective Communication between the Intended Audience and Medical Experts

Lack of solid relationships with potential consumers, stakeholders, and sponsors is a major cause in over 32% of medtech business failures. This shortcoming makes it more difficult for them to get meaningful input to improve and refine their goods. Thus, it is crucial to take into account the following factors when starting a healthcare business: talk to reputable healthcare organisations and specialists to determine any hazards connected to your medical company ideas; get specific input from medical professionals or prospective consumers to make sure your project is headed in the correct direction and effectively solving problems in the real world; keep in close communication with your target market, particularly medical experts, to ensure steady and long-term support for your business.

19.2.3 Risks Related to Development

Other possible obstacles to medical development, such as complicated interfaces, uneven product functionality, and low usability, might also cause medical businesses to fail: technology difficulty or a barrier to tech skills; accessibility of less expensive development options; ineffective marketing and inadequate communication; unreliable healthcare development partner.

19.3 Putting Together the Right Staff for a Healthcare Startup

It is not necessary for an entrepreneur to have in-depth technological knowledge, business planning experience, or healthcare competence in order to succeed in a digital healthcare venture [6]. It is advisable to put up a team with these capabilities instead. Key personnel for a digital healthcare firm include software developers, a chief technology officer (CTO), and the core team.

1. Create a fundamental group: Organise cofounders and important team members by defining essential principles such as a clear company plan, adept management, and a background in healthcare. It's important to have an open mind since your original principles could change as your business expands. Make use of networking sites like LinkedIn, Founders Nation, and CoFoundersLab to make connections with people who are passionate about particular areas. Since your solution is digital, make sure to include a team member with technological skills right away.
2. Hire an experienced CTO: In a company, the CTO plays a crucial role in managing software development, coaching teams, selecting the technology stack, and creating production schedules. This person is crucial when it comes to showing your product to potential customers and investors later on in the startup process. Selecting a CTO with relevant IT experience and expertise in building medical businesses is essential. Look for people with experience, such as senior engineers or software architects who have built successful healthcare solutions in the past. As an alternative, think about employing an outside CTO from a respectable outsourcing company, making sure they are competent by carefully assessing their credentials and abilities.
3. Contract out the creation of software: Creating software for the healthcare industry has intrinsic difficulties beyond creating software for other industries. The most important thing for engineers to understand is the complexities involved in guaranteeing the product's security, compliance, and interoperability. Outsourcing eliminates the need for hiring and training by giving access to a larger talent pool at a more affordable price. Furthermore, 40% of businesses see outsourcing as a way to increase flexibility.

19.4 Guidelines for Establishing a Healthcare Startup

Commencing a medical startup is a time-consuming challenge; hence, the process has been streamlined for optimal industrial solutions. Below outlines an approach to construct a prosperous medical technology startup:

1. Market analysis: Thorough research is crucial in the complex and fiercely competitive healthcare industry. Make a commitment to examining the market, comprehending the intended audience, and investigating current technology. Owing to the complexity of this process, it is best to divide it into many areas:

- Intended audience: The prospective users for your product go beyond patients and may involve healthcare providers like hospitals, laboratories, pharmacies, private practices, medical insurance agencies, and technology firms. It is crucial to precisely identify the individuals and/or entities interested in acquiring your product upon its release.
- Needs and pain spots: It is crucial to recognise the fundamental needs and challenges of the target audience. Furthermore, assessing the significance of these pain points and whether they warrant a solution is essential. This necessitates the team's exploration of medical research papers and conducting interviews with doctors, insurers, and patients.
- Studying market related to healthcare: Another crucial step is to understand the constraints that can keep facilities and patients from using your ideas. This might be due to anything from limited digital literacy in rural areas to sluggish internet- or paper-based procedures.
- Competition in the healthcare market: Examine enterprises that have attempted to address similar challenges, particularly those that were unsuccessful. Additionally, scrutinise indirect competitors who have approached the same issues using different methods.

2. Establishing a Medical Advisory Board is imperative, as a compelling concept alone does not ensure a thriving startup. To achieve success, it is essential to possess a nuanced understanding of medical science and niche intricacies, highlighting the importance of expert insights [5, 7]. The Medical Advisory Board contributes substantial perspectives on your initiatives, identifies potential obstacles, and proposes strategies to fortify your business model. Board members play a vital role in staying informed about impending regulatory changes and technological trends beneficial to your startup. Furthermore, the credibility of experts with a robust reputation in the field carries significant weight with investors, healthcare providers, and other strategic partners. A versatile medical advisory board is advantageous, incorporating a balance of medical subject matter experts and generalists, along with scientists, health software specialists, nurses, and other non-physicians to bring diverse perspectives.
3. Draft the Development Plan: With the help of your development team's experience, create a detailed product blueprint. Goals, expectations, functional and non-functional needs, and links to other systems should all be made clear in the documentation. Important topics to discuss include:

- Software: This group includes application programming interfaces, frameworks, coding languages, and open-source tools.
- Design: This is used to describe the interactive and visual components of a software programme, including the layout and design that users engage with.
- Architecture: This relates to the software system's general architecture and organisation.
- Security: These encompass both the potential threats originating from within the system or organisation and those arising externally.

Transparent documentation aids team members in aligning their efforts, particularly advantageous for startups with distributed teams. Furthermore, it provides clarity for managers to assess the potential costs associated with development.

4. Develop a finance strategy and a business plan: By sustaining the startup for a minimum of five years, it is hoped to avoid running out of resources before debut. This calls for a thorough, long-term strategy that includes goals, deadlines, time estimates, and—most importantly—a budget. Calculating foundational costs is a necessary step in estimating startup costs for the digital health industry. Start with one-time purchases of property, assets, inventories, and security deposits. Next, list the ongoing costs, which should include fixed components such as office rent, insurance, phone and internet services, and bank fees, as well as variable costs like materials, cloud hosting, income taxes, and outsourced services.

   Describe the development costs in accordance with the technical requirements, keeping in mind that costs might go up if more features or mobile versions are added. Validating the concept before putting it into practice is a crucial step in preventing future cost increases.
5. Start with minimum viable product (MVP): It's best to assess the concept, essential features, and business plan before devoting all of your resources to a lengthy development process. An MVP is a streamlined version of the product that has the essential elements. Prioritise important features that answer the audience's main issues; save other features for later releases. Take the creation of a telemedicine platform for hospitals, for instance. The MVP's fundamental components would include video and audio conference functionality, appointment management, and user registration [8]. Following the establishment of these essential features, supplementary elements or API integrations can be introduced.
6. Decide on the revenue model. Startups in the health tech space might choose to use a subscription-based business model or a one-time

licencing charge. One-time payments make sense for wearables, direct-to-patient services, and instruments associated with remote patient monitoring. On the other hand, cloud software vendors usually choose recurring payments since they may draw in more users with lower upfront costs. Most subscription-based services use one of the following price structures:

- Freemium model: A business strategy in which some functions are offered at no cost, but the core features are locked behind a paywall.
- Flat rate pricing: For a set monthly fee, users receive full functionality.
- Pay-as-you-go pricing: Based on consumption indicators unique to each client, the pricing structure is subject to change.
- Based on features and other usage criteria, such as available data, transaction volume, or other pertinent aspects, the solution offers a range of set rates.
- Pricing per user: The number of active users per account decides billing.

Strive to achieve a harmonious balance between costs and perceived value for your target audience. Furthermore, consider experimenting with diverse strategies. A study conducted in 2021 disclosed that 27% of Software as a Service (SaaS) companies adapt their business models by implementing a tiered pricing structure.

7. Verify Standards and Regulation Compliance: Individuals starting a medical business in the healthcare industry need to make sure they follow established data security procedures and comply with local laws. Start by assessing the regulations, administrative framework, and IT infrastructure. Subsequently, identify the personal health information handled, analyse vulnerabilities inside systems, and quantify possible implications of data breaches. Implement technological solutions like data encryption, automated log-off, multi-factor authentication, and intrusion detection systems to reduce the impact of external attacks. Setting up role-based access with authorisation levels specific to each employee's function is also essential.

Ensuring compliance with HIPAA and other regulations entails thorough work, yet you can streamline the task by utilising compliance management tools. These aids help align your software and activities with security standards. However, a more efficient approach involves engaging a seasoned technical company for system audits [9]. A reliable vendor can assist in addressing vulnerabilities in your cybersecurity and provide training to staff on security and compliance practices.

8. Provide your answer to the main target audiences in order to promote your product before it is released. Make sure you are in line with the gatekeepers of the industry, the government and insurance firms. Innovative technology introductions in the medical field might be difficult, necessitating scientific studies to support your idea. Furthermore, committing to putting findings via randomised controlled trials and peer reviews strengthens their legitimacy.
9. Investigate Funding: Business owners starting out have access to a variety of funding sources, including both public and private investors. These investors frequently contribute to the startup's development, marketing, accreditation, mentorship, clinical studies, and development. Typical financing sources for health businesses include:
   - Venture Capital Funds: Only 4% of firms receive money from venture capitalists, according to *The Washington Post.* To add more context, the chance of securing money from a venture capitalist organisation like Andreessen Horowitz is less than 1%. Securing venture capital (VC) money adds more pressure and scrutiny, even in the event of success. These organisations set strict requirements for the businesses they would invest in, including minimum internal rates of return (IRR) and return on investment (ROI). Venture capitalists follow strict deadlines for every round of fundraising. Interestingly, businesses such as Medtronic and Edwards Lifesciences only fund entrepreneurs who want to launch health enterprises.
   - Angel investors: A person who provides funding to businesses in exchange for equity is known as an angel investor. A one-time cash infusion or several instalment payments can be used for financing a startup's early public offering. Famous angel investors in the technology space include Sacha Levy and Naval Ravikant. Use AngelList and AngelMD to find more angel investors for your medical startup.
   - Non-profit funds: Organisations and charities that focus on particular fields of medicine and healthcare may also be able to provide startups with funding. For instance, a startup dedicated to type 1 diabetes research was funded by a partnership between PureTech and the Juvenile Diabetes Research Foundation (JDFR).
   - Accelerators: Small medtech businesses can get funding for their operations with the aid of accelerators. Accelerators might not provide mentoring like venture capitalists do, so you'll have to handle the business and administrative aspects on your own. Under the direction of the Arizona State University Alliance for Health Care and

the Mayo Clinic, MedTech Accelerator is an accelerator programme for startups in the health technology space. Other noteworthy accelerators are Bayer's Medlim and G4A.

- Public funding: To provide money for health research and development, government organisations like the National Institutes of Health (NIH) launch special projects. The National Institute of Mental Health, the National Institute on Aging, and the National Cancer Institute are a few public organisations you can apply to for funding if you want to launch a health technology startup.
- Private funding: Referring back to the previously mentioned *Washington Post* article, over 70% of startup funding comes from private sources (not including charitable organisations). If you have sufficient funds, you can finance your own startup and place a bet on yourself. Personal funding can also come from bank loans, gifts from friends and family, and inheritance.

10. Create and release the finished product: Examine the input you got from the MVP launch and ask your developers to make any necessary changes to the technical requirements before moving forward with the full development. We advise considering scalability when developing your application. Cloud-based infrastructures, which let you customise computing and storage resources to your specifications, can be used for this. Make use of a microservices architecture or modular monolith to separate your software modules into independent, loosely coupled parts. This guarantees that the operation of your system is independent of other parts, making it simple to update modules without having an impact on the system as a whole. You can also add ready-to-use APIs to your software to expand its functionality. Though there is undoubtedly a long road ahead of you, this guide can get you started. But being aware of the traps is also necessary for success.

Entering the healthcare industry as a technology startup can present difficulties in the form of laws and risky initiatives. An entrepreneur can transform revolutionary ideas into workable solutions for humanity and even make money doing it if they take the proper approach. Learn from the mistakes made by unsuccessful startups to create a medical startup that will endure for many years. Utilise data to modify the operational procedures of the healthcare startup. All things considered, if anyone wants a medical startup to succeed, they should always put the needs of people first.

References

1. Aryadita H, Sukoco BM, Lyver M. Founders and the success of start-ups: an integrative review. *Cogent Bus Manag* 2023;10:2284451. https://doi.org/10.1080/23311975.2023.2284451.
2. How to launch a medical startup? n.d. https://appinventiv.com/blog/how-to-build-a-medical-startup/ (accessed January 16, 2024).
3. Thapa RK, Iakovleva T. Responsible innovation in venture creation and firm development: the case of digital innovation in healthcare and welfare services. *J Responsible Innov* 2023;10. https://doi.org/10.1080/23299460.2023.2170624.
4. Singh S, Mungila Hillemane BS. An analysis of the financial performance of tech startups: do quantum and sources of finance make a difference in India? *Small Enterp Res* 2023;30:374–99. https://doi.org/10.1080/13215906.2023.2270447.
5. Blank S, Euchner J. The genesis and future of lean startup: an interview with Steve blank. *Res Technol Manag* 2018;61:15–21. https://doi.org/10.1080/08956308.2018.1495963/ASSET//CMS/ASSET/24F39A17-C235-48AF-8CD6-F7F97CADBCD5/08956308.2018.1495963.FP.PNG.
6. Balasubramanian S, Shukla V, Islam N, Upadhyay A, Duong L. Applying artificial intelligence in healthcare: lessons from the COVID-19 pandemic. *Int J Prod Res* 2023. https://doi.org/10.1080/00207543.2023.2263102.
7. Harms R, Schwery M. Lean startup: operationalizing lean startup capability and testing its performance implications. *J Small Bus Manag* 2020;58:200–23. https://doi.org/10.1080/00472778.2019.1659677.
8. Koskinen H. Outlining startup culture as a global form. *J Cult Econ* 2023. https://doi.org/10.1080/17530350.2023.2216215.
9. Cook DA, Bikkani A, Poterucha Carter MJ. Evaluating education innovations rapidly with build-measure-learn: applying lean startup to health professions education. *Med Teach* 2023;45:167–78. https://doi.org/10.1080/0142159X.2022.2118038.

Index

T

U

V

W

Z

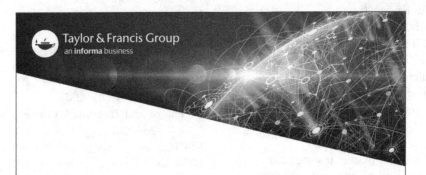

Printed in the United States
by Baker & Taylor Publisher Services

Printed in the United States
by Baker & Taylor Publisher Services